International Explorations in Outdoor and Environmental Education

This series focuses on contemporary trends and issues in outdoor and environmental education, two key fields that are strongly associated with education for sustainability and its associated environmental, social and economic dimensions. It also has an international focus to encourage dialogue across cultures and perspectives. The scope of the series includes formal, nonformal and informal education and the need for different approaches to educational policy and action in the twenty first century.

Research is a particular focus of the volumes, reflecting a diversity of approaches to outdoor and environmental education research and their underlying epistemological and ontological positions through leading edge scholarship.

The scope is also be both global and local, with various volumes exploring the issues arising in different cultural, geographical and political contexts. As such, the series aims to counter the predominantly "white" Western character of current research in both fields and enable cross-cultural and transnational comparisons of educational policy, practice, project development and research.

The purpose of the series is to give voice to leading researchers (and emerging leaders) in these fields from different cultural contexts to stimulate discussion and further research and scholarship to advance the fields through influencing policy and practices in educational settings. The volumes in the series are directed at active and potential researchers and policy makers in the fields. Book proposals for this series may be submitted to the Publishing Editor: Claudia Acuna E-mail: Claudia.Acuna@springer.com

More information about this series at http://www.springer.com/series/11799

Fernando M. Reimers
Editor

Education and Climate Change

The Role of Universities

Springer

Editor
Fernando M. Reimers
Harvard Graduate School of Education
Harvard University
Cambridge, MA, USA

ISSN 2214-4218 ISSN 2214-4226 (electronic)
International Explorations in Outdoor and Environmental Education
ISBN 978-3-030-57926-5 ISBN 978-3-030-57927-2 (eBook)
https://doi.org/10.1007/978-3-030-57927-2

This Springer imprint is published by the registered company Springer Nature Switzerland AG.
The registered company address is: Gewerbestrasse 11, 6330 Cham, Switzerland

Series Editors' Foreword

> Education is an essential element of the global response to climate change. It helps young people understand and address the impact of global warming, encourages changes in their attitudes and behaviour and helps them adapt to climate change-related trends. (UNESCO 2019a, b)

Climate change is probably one of the most contentious global issues of the past 50 years, and climate change education, like other forms of environmental education, has proved to be challenging to implement, so this book is a timely contribution to the catalogue of strategies that can be applied, with a particular focus on both what universities can do and what can be enacted in schools and communities.

In 1966 the World Meteorological Organization used the term "climatic change" to encompass all forms of climatic variability on time-scales longer than 10 years, regardless of whether the cause was natural or anthropogenic. However, when it was realised that human activities had the potential to drastically alter the climate, the term "climate change" became the dominant term to reflect an anthropogenic cause. According to NASA's Erik Conway (2008):

> The first decisive National Academy of Science study of carbon dioxide's impact on climate, published in 1979, abandoned "inadvertent climate modification" … [and] declared: "if carbon dioxide continues to increase, [we find] no reason to doubt that climate changes will result and no reason to believe that these changes will be negligible.

The United Nations Environment Programme (UNEP) and the World Meteorological Organization set up the International Panel on Climate Change (IPCC) in 1988, but it was not until 1994 (as an outcome from the 1992 Earth Summit held in Rio de Janeiro) that the UN Framework Convention on Climate Change (UNFCCC) came into force. The 2015 Paris Agreement builds upon the Convention and brings together 189 signatory nations into a common cause to undertake ambitious efforts to combat climate change and adapt to its effects. It also complements the United Nations (2016a) Sustainable Development Goals, which includes the need for climate action, because "Climate change is a global challenge that affects everyone, everywhere" (United Nations 2016a), as one of the 17 Sustainable Development Goals. This goal states: "Take urgent action to combat climate change and its impacts" (United Nations 2016b). Of particular relevance within this goal is Target 13.3: "Improve education, awareness-raising and human and institutional capacity on climate change mitigation, adaptation, impact reduction and early warning". (United Nations 2016b).

According to the Paris Agreement (UNFCCC 2015, pp. 1–2):

> *Acknowledging* that climate change is a common concern of humankind, Parties should, when taking action to address climate change, respect, promote and consider their respective obligations on human rights, the right to health, the rights of indigenous peoples, local communities, migrants, children, persons with disabilities and people in vulnerable situations and the right to development, as well as gender equality, empowerment of women and intergenerational equity…
>
> *Affirming* the importance of education, training, public awareness, public participation, public access to information and cooperation at all levels on the matters addressed in this Agreement.

Despite this high-level recognition of the need to combat climate change, the notion of climate change has generated more sceptics than virtually any other environmental problem that has been identified, and, unfortunately, this scepticism seems to have spread to the ways that many sectors of society, especially government, now treat scientific findings related to all environmental problems. In the current climate, the scientists' messages have been marginalised and now often fall on deaf ears, particularly with Western politicians, but also in the education sectors of most countries.

Climate change education is related to both environmental education and disaster risk reduction and resilience education, two areas that have struggled to find a place in the curriculum and practices of education institutions, even though environmental education has been on the international education agenda since the 1972 United Nations Conference on the Human Environment. Disaster risk reduction and resilience education is a more recent field, but it has risen to prominence through the efforts of the United Nations since the 1994 World Conference on Natural Disaster Reduction. The current guiding document is the *Sendai Framework for Disaster Risk Reduction 2015–2030* (United Nations 2015) and this is complemented by the United Nations Office for Disaster Risk Reduction's *Comprehensive School Safety* framework (GADRRRES 2017).

Climate change is a natural hazard around the globe, and in some of the cases discussed in this volume (particularly Haiti and Pakistan), natural hazard has become a disaster and we need education related to disaster risk reduction and resilience for the populations. As Jo Scheuer (2012) explains,

> A natural hazard only becomes a disaster when measures to mitigate its impact, such as earthquake resistant buildings, are lacking. We don't have to resign ourselves to the devastation that disasters cause, nor see them as exceptional events that interrupt normal development.

Indeed, those who work in disaster risk reduction see "natural disaster" as a harmful expression because

> it facilitates the concealment of the human, political and social responsibility that exists behind every disaster caused by natural or technological threats. It confounds threats with

> risk and discriminates against vulnerability… Disasters are not natural; they are products of historically determined social processes. (Global Network of Civil Society Organisations for Disaster Reduction (GNDR) 2015).

Put this way, climate change is very much a natural hazard, and as the stories in Chaps. 2, 3, 4 and 5 describe, this natural hazard has become a disaster as a result of historically determined social processes.

Universities around the world have an important role to play in climate change education, and this volume describes the efforts of students in one postgraduate program at Harvard University to implement climate change education across formal and nonformal education settings in five different countries.

Some universities moved to make a commitment to environmental sustainability 30 years ago. The Talloires Declaration was formulated in 1990 by, what became, the Association of Universities Leaders for a Sustainable Future at a meeting in Talloires, France. This Declaration provides a 10-point action plan for incorporating sustainability and environmental literacy in teaching, research, operations and community responsibilities in universities. Originally signed by 31 university representatives, there are now over 500 signatories from over 50 countries. Harvard is not a signatory to the Talloires Declaration, but it is a charter member of the smaller International Sustainable Campus Network (ISCN) which has as its mission "to provide an international forum to support higher education institutions in the exchange of information, ideas, and best practices for achieving sustainable campus operations and integrating sustainability in research and teaching" (ISCN 2019). Other international networks focused on implementing strategies for the incorporation of sustainable development in universities and other higher education institutions include the COPERNICUS Alliance and the Global Higher Education for Sustainability Partnership (GHESP). Unfortunately, not all universities are yet committed to environmental sustainability or the Sustainable Development Goals (United Nations 2016a) and are more consistent with what Ray Anderson commented on nearly 20 years ago,

> The educational system, along with society in general, still treats the Earth as if it's an infinite resource. The universities – in their academic programs, credit requirements, curricula, course design, campus design, and campus operation – perpetuate this flawed mind-set from generation to generation, with scarcely a pang of conscience, much less a serious re-examination of [their] role in the destruction of the biosphere (Anderson, quoted in Wakefield, 2003, p. 5).

Fernando Reimers and his students hope that the stories in his volume will help to encourage universities to embrace sustainability and, particularly, climate change education. Their stories of implementing climate change education in the different countries and contexts are taking up many of the challenges that Naomi Klein argued in her 2016 Edward W. Said London lecture:

> Climate change acts as an accelerant to many of our social ills – inequality, wars, racism – but it can also be an accelerant for the opposite, for the forces working for economic and social justice and against militarism. Indeed the climate crisis – by presenting our species with an existential threat and putting us on a firm and unyielding science-based deadline – might just be the catalyst we need to knit together a great many powerful movements,

bound together by a belief in the inherent worth and value of all people and united by a rejection of the sacrifice zone mentality, whether it applies to peoples or places. We face so many overlapping and intersecting crises that we can't afford to fix them one at a time. We need integrated solutions, solutions that radically bring down emissions, while creating huge numbers of good, unionised jobs and delivering meaningful justice to those who have been most abused and excluded under the current extractive economy (n.p.).

It was part of the process of policy analysis that the Harvard students were learning in the course that they had to identify how specific impacts of climate change in their chosen locality affected various human populations and to identify which of these populations had to be educated on climate change. Sadly, none of the students identified women as a target group. However, as Greta Gaard (2015, p. 23) highlights, "Make no mistake: women are indeed the ones most severely affected by climate change and natural disasters[sic], but their vulnerability is not innate; rather it is a result of inequities produced through gendered social roles, discrimination, and poverty". According to the Global Gender and Climate Alliance (2013):

- Women in developing countries are particularly vulnerable to climate change because they are highly dependent on natural resources for their livelihood.
- Women experience unequal access to resources and decision-making processes, with limited mobility in rural areas.
- Women make between 30% and 80% of what men earn annually.
- 103 out of 140 countries surveyed by the World Bank impose legal differences on the basis of gender that may hinder women's economic opportunities.
- Women make up half of the agricultural workforce in the least developed countries.
- In developing countries women own between 10% and 20% of the land.
- Two thirds of the world's illiterate adults are women.
- Socio-economic norms can limit women from acquiring the information and skills necessary to escape or avoid hazards (e.g. swimming or climbing trees to escape rising water levels).
- Dress codes imposed on women can restrict their mobility in times of disaster, as can their responsibility for small children who cannot swim or run.
- A lack of sex disaggregated data in all sectors often leads to an underestimation of women's roles and contributions, thus increasing gender-based vulnerability.

Perhaps some of these points led to the invisibility of women in the case studies.

The realities of climate change hit home in the past summer in Australia. Not only was the country in the midst of one of the worst droughts on record, the bushfire fire season in Australia in 2019 started months earlier than previously and continued into March 2020, a period that is now called the Black Summer. These fires changed the way many Australians think of climate change (Abram et al. 2020; Flanagan 2020; Marshall 2020). That the country (and its oceans) is getting hotter is not in doubt, except to the climate change sceptics, but even they cannot deny that we had the three hottest December days ever recorded in 2019 (Morton 2019). The drought continues over much of the continent, and in 2020, the Great Barrier Reef had its most widespread bleaching event to date. Yet, Australian politicians, like those in many other countries, continue to refuse to act on climate change and indeed continue to pursue a fossil fuelled future agenda (Irfan 2020), and are increasingly marginalising climate change in the curriculum (Gough 2020). The ramifications of this are enormous – not just for Australia but for the planet. As Josie Douglas from the Central Land Council in Mparntwe (Alice Springs) said recently, "Without action to stop climate change, people will be forced to leave their country and leave behind much of what makes them Aboriginal. Climate change is a clear and present threat to the survival of our people and their culture". (in Allam and Evershed 2019).

It is important to note that Australia is not alone in having a government that is resisting action on climate change. When Theresa May was appointed British Prime Minister in 2016, one of her first acts was to abolish the Department for Energy and Climate Change and move responsibility for climate change to a new Department for Business, Energy & Industrial Strategy. This situation continues today.

> Climate change is a real and rapidly-evolving threat for humanity, and in the words of the United Nations Secretary-General, it is "a battle for our lives". It is one of the focus areas of ESD, through which UNESCO strives to ensure that all generations understand the impact of climate change and are better equipped to take action to protect resources, the environment and the planet that sustains life, as enshrined in SDG 13. These actions also support Member States in meeting their obligations under the United Nations Framework Convention on Climate Change and 2015 Paris Agreement by advocating on the importance of education as a key element of the response to climate change; producing and sharing knowledge and policy guidance on climate change education; and providing country support and in the implementation of projects on the ground. (UNESCO 2019a, b, p. 2)

We need universities and schools and community organisations as well as politicians and industry to be acting responsibly to combat climate change. "Our house is on fire" proclaimed Greta Thunberg (2019), and following her, thousands of young people around the world protested governmental inaction on climate change. We also need to explore different pedagogies for climate change education, such as Blanche Verlie's and CCR15's (2018) on becoming-with climate change. We hope that the readers of this volume will follow in their stead and amplify their efforts in implementing climate change education.

RMIT University
Melbourne, VIC, Australia
Annette Gough

La Trobe University
Melbourne, VIC, Australia
Noel Gough

References

Abram, N. et al. (2020, 3 February). *There is no strong, resilient Australia without deep cuts to greenhouse gas emissions*. Retrieved from https://australianbushfiresandclimatechange.com/

Allam, L., & Evershed, N. (2019, 18 December). Too hot for humans? First Nations people fear becoming Australia's first climate refugees. *The Guardian*. https://www.theguardian.com/australia-news/2019/dec/18/too-hot-for-humans-first-nations-people-fear-becoming-australias-first-climate-refugees?CMP=Share_iOSApp_Other

Conway, E. (2008). *What's in a name? Global warming vs climate change*. Retrieved from www.nasa.gov/topics/earth/features/climate_by_any_other_name.html

Flanagan, R. (2020, 25 January). How does a nation adapt to its own murder? *The New York Times*. Retrieved from https://www.nytimes.com/2020/01/25/opinion/sunday/australia-fires-climate-change.html

Gaard, G. (2015). Ecofeminism and climate change. *Women's Studies International Forum, 49*, 20–33.

Global Alliance for Disaster Risk Reduction & Resilience in the Education Sector (GADRRRES) and United Nations Office for Disaster Risk Reduction (UNIDSR). (2017). *Comprehensive School Safety*. Retrieved from https://gadrrres.net/resources/comprehensive-school-safety-framework

Global Gender and Climate Alliance (GGCA). (2013). *Overview of linkages between gender and climate change*. Retrieved from https://issuu.com/undp/docs/pb1_asiapacific_capacity_final

Global Network of Civil Society Organisations for Disaster Reduction (GNDR). (2015, 14 September). *Disasters are not natural!* https://www.gndr.org/news/item/1499-disasters-are-not-natural.html

Gough, A. (2020). Educating Australia on the climate crisis. *Policy Forum*. https://www.policyforum.net/educating-australia-on-the-climate-crisis

International Sustainable Campus Network. (2019). *Members*. https://international-sustainable-campus-network.org/members/

Irfan, U. (2020, 8 January). As Australia burns, its leaders are clinging to coal. *Vox*. https://www.vox.com/world/2020/1/8/21051756/australia-fires-climate-change-coal-politics

Keneally, T. (2020, 1 February). Thomas Keneally: 'These fires have changed us'. *The Guardian*. Retrieved from https://www.theguardian.com/australia-news/2020/feb/01/thomas-keneally-these-fires-have-changed-us?CMP=share_btn_link

Klein, N. (2016, 2 June). Let Them Drown: The Violence of Othering in a Warming World. *London Review of Books, 38*(11). Retrieved from www.lrb.co.uk/v38/n11/naomi-klein/let-them-drown

Marshall, K. (2020, 25 January). The 'Forever fires' and Australia's new reality. *The Age Good Weekend*. Retrieved from https://www.theage.com.au/national/the-forever-fires-and-australia-s-new-reality-20200122-p53tk0.html

Morton, A. (2019, 21 December). Climate of chaos: The suffocating firestorm engulfing Australia. *The Guardian*. Retrieved from https://www.theguardian.com/australia-news/2019/dec/20/climate-of-chaos-the-suffocating-firestorm-engulfing-australia

Rosane, O. (2020, 7 April). 2020 Great Barrier Reef Bleaching Event Is Most Widespread to Date. *EcoWatch*. Retrieved from https://www.ecowatch.com/great-barrier-reef-bleaching-2020-2645655957.html?rebelltitem=2#rebelltitem2

Scheuer, J. (2012, 12 October). *What we call "natural" disasters are not natural at all*. Retrieved from https://www.undp.org/content/undp/en/home/ourperspective/ourperspectivearticles/2012/10/12/what-we-call-natural-disasters-are-not-natural-at-all-jo-scheuer.html

Thunberg, G. (2019, 29 January). 'Our house is on fire': Greta Thunberg, 16, urges leaders to act on climate. *The Guardian*. Retrieved from https://www.theguardian.com/environment/2019/jan/25/our-house-is-on-fire-greta-thunberg16-urges-leaders-to-act-on-climate

UNESCO. (2019a). *Climate change education*. https://en.unesco.org/themes/education-sustainable-development/cce

UNESCO. (2019b). *Framework for the Implementation of Education for Sustainable Development Beyond 2019*. https://unesdoc.unesco.org/ark:/48223/pf0000370215.page=7

United Nations. (1994). *Yokohama strategy for a safer world: Guidelines for natural disaster prevention, preparedness and mitigation.* https://www.preventionweb.net/files/8241_doc6841contenido1.pdf

United Nations. (2015). *United Nations Sendai Framework for Disaster Risk Reduction 2015–2030.* https://www.unisdr.org/files/43291_sendaiframeworkfordrren.pdf

United Nations. (2016a). *About the sustainable development goals.* Retrieved from https://www.un.org/sustainabledevelopment/sustainable-development-goals/

United Nations. (2016b). *Goal 13: Take urgent action to combat climate change and its impacts.* Retrieved from https://www.un.org/sustainabledevelopment/climate-change/

United Nations Framework Convention on Climate Change (UNFCCC). (2015). *The Paris agreement.* Retrieved from https://unfccc.int/files/essential_background/convention/application/pdf/english_paris_agreement.pdf

Verlie, B., & CCR15. (2018). From action to intra-action? Agency, identity and 'goals' in a relational approach to climate change education. *Environmental Education Research*, 1–15. https://doi.org/10.1080/13504622.2018.1497147

Wakefield, J. (2003). Teaching sustainability. *Environmental Health Perspectives, 111*(5), 270.

Contents

Chapter 1
The Role of Universities Building an Ecosystem of Climate Change Education

Fernando M. Reimers

1.1 Introduction. The Paradox of Climate Change and Education

Along with many species on the planet, polar bears are experiencing the effects of climate change. As rising temperatures produce ice loss, the bears lose the platforms they use to hunt seals. This change to their habitat has placed polar bears on the list of species on the risk of extinction in the wild (Pidcock 2015). The bear floating on a melting ice platform has become an icon for the potentially devastating consequences of climate change to life on the planet.

Like bears, humans also face risks to their habitat and survival resulting from climate change, such as the intensification of the wildfire season ravaging the West Coast in the United States, or more intense storms, droughts and floods, rising sea levels or increasing temperatures. Unlike bears, however, humans have invented an institution to help us quickly adapt to and mitigate changing trends: schools. Furthermore, because climate change is largely the result of human-environmental interactions (IPCC 2018, p 53), schools can do more than help us understand these changes to our habitat, or help us adapt to those changes, they could help us slow down those changes and mitigate their impact, as we adopt practices that are more sustainable, and perhaps even revert them, as we invent technologies that transform the drivers of climate change.

The International Panel on Climate Change (IPCC), has established that human activities have caused global temperature to rise about 1.0 °C degrees above preindustrial levels (before the 1880s) and that, if increases continue on the current trajectory, global warming will likely reach 1.5 °C between 2030 and 2052. These changes to climate pose risks to health, livelihoods, food security, water supply, human security, and economic growth, and will increase as temperatures reach

F. M. Reimers (✉)
Harvard Graduate School of Education, Harvard University, Cambridge, MA, USA
e-mail: Fernando_Reimers@harvard.edu

F. M. Reimers (ed.), *Education and Climate Change*, International Explorations in Outdoor and Environmental Education,
https://doi.org/10.1007/978-3-030-57927-2_1

1.5 °C and increase further as temperature increase to 2 °C above preindustrial levels. The impact of global warming of 1.5 °C and beyond will be greater on disadvantaged and vulnerable populations, indigenous peoples, and communities whose livelihood is dependent on agricultural or coastal activities. Global warming will also contribute to increases in poverty (IPCC 2018, p. 4–9.) and will have a disproportionate impact on women who are poor and from other disadvantaged groups and whose livelihood depends largely on agriculture.

> "Differences in vulnerability and exposure arise from non-climatic factors and from multidimensional inequalities often produced by uneven development processes (very high confidence). These differences shape differential risks from climate change. People who are socially, economically, culturally, politically, institutionally or otherwise marginalized are especially vulnerable to climate change and also to some adaptation and mitigation responses (medium evidence, high agreement). This heightened vulnerability is rarely due to a single cause. Rather, it is the product of intersecting social processes that result in inequalities in socio-economic status and income, as well as in exposure. Such social processes include, for example, discrimination on the basis of gender, class, ethnicity, age and (dis)ability." (IPCC 2014, p 54).

The IPCC identifies a range of education options to adapt to and mitigate climate change, including awareness raising and integration of climate change education in school curricula, gender equity in education; and various forms of adult and non-formal education, including extension services; sharing indigenous, traditional & local knowledge; participatory action research & social learning; Knowledge-sharing & learning platforms and disseminating information on hazards and vulnerability (IPCC 2014, p.27).

Educating people for more sustainable ways to relate to our habitat involves preparing us to adopt sustainable practices that reduce our impact on climate change and the impact of climate change in our lives. These practices may be individual, in the choices we make about our own consumption and lifestyle (for example slowing down population growth, consuming a diet with a smaller carbon footprint or using renewable energies, or consuming less), or they may be collective, the result of choices we make as citizens when we participate in the democratic process at various levels of government, our towns or cities, states, or nations, or when we influence the behavior of corporations (for example adopting caps to emissions or a carbon tax, or incentivizing the reliance on clean energies). Government policies such as caps on emissions are essential to slowing global warming, and they are subject to influence and preferences by citizens, educated to understand the scientific consensus on climate change and with the capacity to exercise influence as citizens. Collective responses may also include shaping the way in which we live and our habitats, for instance the value we assign to nature as we design and build the homes and cities where we live and work.

In addition to personal responsibility for our individual impact on climate change, and participation in collective processes that support systemic changes in the norms and institutions that undergird climate change, slowing down, and perhaps over time reverting, climate change requires also advancing knowledge and inventing technologies that can help us transform our interactions with the environment, in a way helping us reinvent our way of life, and so educating for sustainability involves equipping people with the ethical frameworks, the imagination and the necessary skills for such advancement of knowledge and invention. Example of such design and invention and changes to our way of life include developing a circular economy with production of goods next to cities to reduce transportation costs, as well as urbanization with populations concentrated in sustainable cities, or geoengineering the atmosphere to partially block the sun's rays.

An example from the field of sanitation will illustrate the value of technological invention to address climate change. In his efforts to improve health and sanitation in the developing world, Bill Gates concluded that the toilets and water treatment systems developed and in use in the early industrialized world were inadequate to improving sanitation in developing countries because they were resource intensive, generated excessive waste and required intricate and expensive sewer systems to operate. As a result, as an approach to dispose of human waste, toilets are likely to remain out of the reach of a significant share of the world's population. This caused him to undertake projects to stimulate innovation in the design of next-generation toilets that could operate without sewer systems and that could be extended to all of humanity within a relatively short period of time (Brueck 2019; D'Agostino 2018). Similar technological breakthroughs could change our dependence on fossil fuels, help us produce much safer nuclear energy, increase the efficiency of fossil fuels and of clean energies. But it is not just technological advancements that can help us reinvent a way of life, inventions in how we organize our lives and work, and in how we organize our communities can help us mitigate and adapt to climate change. For example, structuring some workplaces in ways that allows working from home can reduce our consumption of fuels. Ethical and spiritual development can stimulate such social innovation and lead us to make different choices placing different value on individual consumption relative to protection of the environment, other forms of life or cause us to seek greater balance across a range of goals in the communities of which we are a part. The United Nations Sustainable Development Goals, for example, are a framework of seventeen interdependent goals that aim at producing a world that is more inclusive and sustainable, and they can provide a normative framework to guide the development of communities, cities, or other jurisdictions.

The motivation to invent more sustainable ways of life requires more than an understanding of the science of climate change, the capacity to design technological innovations, or an ethical framework that help us aspire to live in more inclusive and sustainable communities, it requires an understanding of social systems and the development of ethical reasoning that can help us integrate critical thinking about the current impact of climate change, our moral imagination, the personal motivation to act and our competency to act in effective ways. An example of the integration of understanding of complex social systems with ethical reasoning would be engaging students in projects that helped them understand the gendered experience of climate change. A number of reports explain that the dependence of women in developing countries on natural resources makes them particularly vulnerable to climate change. The challenges are greater for women who secure water, food and fuel for cooking and heating and whose livelihood depends on agriculture. A number of studies of the gendered nature of climate change argue that these differences are the result of various intersectionalities that place particular groups of women (poor, lower casts) at greater risk (Arora-Jonsson 2011). Therefore, understanding intersectionality is necessary to better understand the gendered impact of climate change. The drivers of these gender differences include disparities in access to education, use of time, access to credit and markets, recognition of rights within legal frameworks, and resulting disparities in earnings (UNDP 2013) so understanding these drivers requires understanding systems and complex causality. The recognition of the gendered impact of climate change is the foundation for the recognition of the co-benefits between gender equality and climate action (UN Women 2016) this provides an opportunity for students to understand deeply complex social action and how "equity, sustainable development, and poverty eradication are best understood as mutually supportive and co-achievable within the context of climate action and are underpinned by various other international hard and soft law instruments" (IPCC 2018, p 54).

Developing the moral imagination of students through Human Rights education or education for social justice, cultivating their capacity to recognize how the impact of climate change varies for different people (women, minorities, the poor) is a necessary step to animating them to engage with the subject at greater levels of complexity and inventiveness.

As illustrated with the previous discussion of the complementarities between gender equity and climate change, given the multidimensional nature of the impacts of climate change underscored in recent reports of the IPCC, effective collective responses require addressing the systems that undergird such multidimensional processes. This understanding has led to a growing realization that climate action is best undertaken in coordination in the context of poverty reduction and sustainability efforts, such as those reflected in the development compact adopted at the UN 2015 General Assembly: the Sustainable Development Goals. Advancing such systemic multidimensional efforts requires that we educate students to understand systemic complexity, and develop their capacity to collaborate with others to influence social systems.

> "Differences in vulnerability and exposure arise from non-climatic factors and from multidimensional inequalities often produced by uneven development processes (very high confidence). These differences shape differential risks from climate change. People who are socially, economically, culturally, politically, institutionally or otherwise marginalized are especially vulnerable to climate change and also to some adaptation and mitigation responses (medium evidence, high agreement). This heightened vulnerability is rarely due to a single cause. Rather, it is the product of intersecting social processes that result in inequalities in socio-economic status and income, as well as in exposure. Such social processes include, for example, discrimination on the basis of gender, class, ethnicity, age and (dis)ability." (IPCC 2014, p 54).

To sum up, human competencies, the knowledge, motivation and skills of people, are critical to adapting to and mitigating climate change. Developing those competencies is, however, a very tall order, one that requires focus and specialization.

This potential of education to affect human-environmental interactions has given rise to a new educational domain of education: climate change education, a subfield of education for sustainable development. Much has been written, and is being done, to educate students to understand, adapt to and mitigate climate change. Governments around the world, in partnership with organizations of civil society and with other institutions, have developed climate change curriculum and adopted policies to address this serious risk faced by humanity. International organizations, UNESCO in particular, have advocated extensively for climate change education and developed and distributed resources to support it. In spite of these efforts, education has not yet sufficiently curbed the impact of our own species on climate change, nor have we yet adapted to these climatic changes and as a result, like polar bears, we are witnessing the destruction of our habitat, much of such destruction of our own doing, and wondering whether we will, along with other species, survive such changes.

> "Short of some technological revolution that would transform global energy use, we should be concerned, even alarmed, about the future impact of climate change on the world. It is the quintessential global challenge in that no single country can solve this problem on its own and there is no way for any single country to shield itself from its effects. Generating the required collective response, however, seems highly unlikely. As a result, climate change could conceivably be the defining issue of this century." (Haas 2020, p. 192).

1.2 Climate Is Changing Faster Than Attitudes and Behaviors About Human-Environmental Interactions, and Knowledge Is Not Enough to Cause People to Adapt or Mitigate

While there is arguably more interest around the world in environmental sustainability, and in climate change, than at any time in human history (Mayherfeld and Askhood 2015), it is also the case that our climate challenges are greater than ever. The question then is not whether the environmental movement has increased awareness and action to address climate change, the question is whether it has done so on a scale and level of effectiveness commensurate with the nature of our present challenge and with the velocity at which the challenge is augmenting.

A recent survey of sustainable development experts and practitioners from business, government, NGOs and academia reveals that more than half of those surveyed believe that the rate of progress with respect to climate change is insufficient to avert major damage to human, social and ecosystem health, and less than a third of them believe that good progress is being made implementing the global framework adopted in the Paris agreement (GlobeScan 2017, p. 4). Furthermore, climate optimism (the belief that society is making progress fast enough to avert major irreversible damage to human, social and ecosystem health) has declined considerably over the last fifteen years. In North America, close to 20% of respondents believed it likely or very likely that we were making adequate progress in 2003, compared with 11% in 2017. Climate optimism, has also declined for the rest of the world, from 11% in 2003 to 5% in 2017 (GlobeScan 2017, p. 11).

The most recent World Economic Forum Report on Global Risks identifies climate related risks as the most likely of all risks humanity faces. They include extreme weather, climate action failure, natural hazards, biodiversity loss and human made environmental disasters. Of those, climate action failure, biodiversity loss and extreme weather are also among the five most potentially impactful risks (World Economic Forum 2020).

Climate change, observable changes in climate patterns resulting from a warming of the temperature of the atmosphere, is the result of human activity, principally burning of fossil fuels which release carbon dioxide and other greenhouse gases into the atmosphere where they trap the sun's rays, thereby increasing temperature. The five warmest years since 1880 have all occurred since 2015 (NOAA 2020). Climate change is causing increases in sea levels in coastal areas, more severe storms, higher temperatures and increasing desertification and wildfires. This will reduce productive land. Rising temperatures and salinization will endanger many life forms, crop yields and disease prevalence. Freshwater shortages, extreme heat, flooding and storms will cause large scale migration. The two main drivers of climate change are consumption of fossil fuels and deforestation (Haas 2020, pp. 183–186).

Increases in consumption of fossil fuels stem from the growing levels of consumption and waste produced by a growing population and by the energy and waste produced by modern manufacturing. Consumption, population growth, energy and waste are therefore the major drivers of climate change producing significant release of carbon dioxide and other bases into the atmosphere which trap heat. The resulting warming of the planet is causing rapid and extensive biodiversity loss and land degradation (UNEP 2012; UNESCO 2016). Over the last fifty years fossil fuel consumption has tripled, largely a result of increases in transportation, construction and industrial manufacturing (Haas 2020, p. 185).

Scientists have identified boundaries for ten systems that affect life for humans and other species: freshwater use, land use, phosphorus pollution, ocean acidification, climate change, ozone depletion, nitrogen pollution, biodiversity loss, aerosols and chemical pollution. While we have no data on how aerosols and chemical pollution has changed since preindustrial levels, for eight of those system metrics for which we do have data to compare pre-industrial revolution levels to current levels, five of them exceed the boundaries representing high risk that life is not sustainable. These systems are: ocean acidification, climate change, ozone depletion, nitrogen pollution, biodiversity loss. Furthermore, the remaining three metrics: freshwater use, land use and phosphorus pollution, have changed significantly, in the direction of the increasing risk boundary. Only two of the eight metrics (ocean acidification and ozone depletion) have current values which are lower than the values before the industrial revolution, although they remain above the proposed boundary representing high risk (UNESCO 2016, p. 20). The most commonly accepted explanations for those changes focus on overpopulation, modern lifestyles and individual behavior (UNESCO 2016), as well as industrialization and the release of carbon dioxide in the atmosphere.

Population growth, and the energy and resources that more people consume, is a major driver of climate change. It took one thousand years for the world population to grow from an estimated 190,000,000 in the year 200 to 360,000,000 in the year 1200, and another six hundred years to reach one billion in the year 1804. But the improvement in life expectancy associated with medical and public health developments grew the world population by an additional billion people in barely a century. Then, in less than four decades, world population grew from 2 billion in 1927 to over 3 billion in 1960. The next additional billion in world population took only fourteen years, exceeding four billion in 1974. The next billion humans took only thirteen years, the next billion twelve years, and the next additional billion another twelve years, until the world population reached more than seven billion by 2011 (Worldometers 2019). Such exponential growth in the number of humans making demands on the planet is a driver for the changes to the environmental systems discussed earlier. Some of those demands on natural resources and on the atmosphere are the product not just of the number of people but of particular forms of consumption, of forms of life and social and economic organization. For instance, construction, transportation and manufacturing account for most of our consumption of fossil fuels. Circular economies and alternative forms of urbanization can reduce these costs of transportation and construction.

Providing individuals with access to jobs and income, so they can sustain those forms of consumption, and the necessity to grow aggregate economic output to expand such access in the face of a growing population, has been a widely accepted view of 'progress' for many nations for much of the twentieth century. Only during the last decades of the century did the notion that there may be tradeoffs between jobs and consumption, or economic growth and sustainability, become increasingly accepted among government leaders and development professionals, this shift was significantly aided by the United Nations Conference on Environment and Development (Earth Summit) which took place in June 1992 in Rio de Janeiro, Brazil (UNCED 1992). The question of what constitutes development, and how to integrate sustainable development with other social and economic goals, such as poverty reduction and improvement in material standards of living, is at the heart of today's conversation about what is sustainable development. As with climate change, if such conversation is to have any consequence on the ten systems which are moving the planet towards levels of increased risk for life, it needs to move beyond a conversation among elites in governments and international organizations, and become part of how most humans on the planet think about what is a good life and how they make individual and collective choices that lead to greater sustainability. For instance, governments could agree on a total amount of emissions or on a tax of emissions, something no government has yet endorsed (Haas 2020, p. 189). Similarly, individuals could choose to protect natural habitats and consume less, including reducing their living space, or change their diet, as a way to reduce their carbon footprint.

Climate change education must equip most humans with the knowledge, critical thinking skills, understanding of science, and ethical frameworks, that helps them with mitigation, adaptation and reversal of climate change. Mitigation involves trying to slow down the rate of climate change. Adaptation involves reducing the impact of climate change on people. Reversal is a nascent area involving geoengineering, such as planning particles in the atmosphere to partially block sun's rays (Haas 2020, p. 191).

This means that climate change education should go beyond equipping people with the skills to understand climate change, important as that is. It must equip them to understand tradeoffs, to make choices and to invent solutions that can help us integrate choices that are environmentally sustainable within a larger framework of how we live. A simple way to represent the choices involved in mitigation would be to ask how much people are willing to give up, or to pay, or to change their lifestyle, to reduce their own carbon footprint, for example demanding that their governments adopt cap and trade that limit the total emissions or a carbon tax. Educating people for adaptation involves knowledge and skills to change their lifestyles in ways that are responsive to the impact of climate, for example influencing how they build or rebuild their cities so they are more resilient to coastal flooding. Finally, educating them for climate change reversal requires developing the talent and ingenuity that can lead to technological innovations. This means education not just to understand, but to act effectively, in influencing not just personal patterns of consumption, but with the agency and efficacy to collaborate with others to influence the complex systems which undergird climate change.

Awareness and knowledge about climate change appear to be insufficient because while there is evidence that a significant percentage of people are already aware of environmental degradation, and consider it a challenge, climate continues to change. The World Values Survey, a project coordinated at the University of Michigan, collects data from representative samples of the population in a large number of countries on a range of issues. When people were asked to identify the most important world problem, environmental degradation is considered one of the most important problems in the world, second only to poverty which is considered the most important problem by most people in most countries. Environmental pollution is considered the most important world problem in more countries than discrimination against women, and also more important than poor sanitation and infectious diseases. Only with respect to poor education is the number of countries that see this as the most important problem in the world similar to those who see environmental degradation as the most important problem. The results can be seen in Table 1.1.

However, even though people are aware of the importance of environmental sustainability, there is less evidence that they have the skills to translate such awareness into actions that contribute to more sustainable ways of living. This is because human behavior with respect to relating to the environment is not just a function of what we know about that relationship, but about how we weigh the tradeoffs involved in relating in different ways. People may be aware of the fact that walking or riding a bicycle have a smaller carbon footprint than using other forms of transport, and still prefer the convenience, or the income they could draw from the time saved, associated with public transportation or even private transportation. Few workplaces allow extensive telecommuting for work even though doing so would reduce the carbon impact of transportation, and of office space. There are today, effective alternatives to air travel, in the form of highly effective telecommunication

Table 1.1 Percentage of representative samples of adults who see the following as the most serious problem in the world 2010–2014

	People living in poverty and need	Discrimination against girls and women	Poor sanitation and infectious diseases	Inadequate education	Environmental pollution
Total	56.5	7.8	10.2	11.6	12.7
Algeria	50.7	9.6	16.6	8.8	10.9
Azerbaijan	52.8	11.7	11.3	15.6	7.8
Argentina	59.3	8.3	3.6	18.8	9.1
Australia	61.6	4.2	10.3	12.1	11.5
Armenia	74.7	2.1	7.1	3.1	10.6
Brazil	59.2	10.1	10.5	14	5.8
Belarus	61.5	3.3	10.3	3.1	21.3
Chile	61.2	5.9	5.5	21	5.8
China	42.5	5.1	8.5	12.1	21.6
Taiwan	39.9	3.5	10.4	6.1	36.2
Colombia	56.8	14	2.1	13.6	13.5
Cyprus	57.3	6.3	18.7	5.8	11.9
Ecuador	43	8.8	6.9	18	23.1
Estonia	60.3	2.8	10	5.9	20.2
Georgia	73.1	5.2	8.2	4.7	7.7
Palestine	68.9	5.4	6.7	8.4	9.7
Germany	55.8	7.6	6.5	19.3	10.3
Ghana	58.5	6	14.2	16.8	4.5
Haiti	55.5	15.1	7.8	18.8	1.7
Hong Kong	35.3	5	21.6	12.1	25.6
India	58.5	15	4	14.7	7.5
Iraq	60.8	5.8	7.5	12.8	12.2
Japan	36.6	0.7	7.2	8.1	41.3
Kazakhstan	62.6	3.4	11.8	8	14.2
Jordan	80.7	2.5	5.2	7	4.2
South Korea	42.3	6.3	7.4	4	39.5
Kuwait	49.1	5.4	13.3	21.7	7.3
Kyrgyzstan	54	11.6	13.4	8.7	12.3
Lebanon	48	7.2	17	15.2	11.5
Libya	29.2	3.7	28	26.5	11
Malaysia	44.1	10.9	12.4	16.2	16.5
Mexico	44.4	14.1	5	15.4	20.3
Morocco	62	5.6	6.5	17.6	6.9
Netherlands	64.2	11.6	7.8	5.6	10.3
New Zealand	53.4	3.2	9	15.5	14
Nigeria	79.9	8.2	5.3	5.4	1.2
Pakistan	62.6	20.5	7.8	8.6	0.4

(continued)

Table 1.1 (continued)

	People living in poverty and need	Discrimination against girls and women	Poor sanitation and infectious diseases	Inadequate education	Environmental pollution
Peru	48	7.4	3.8	22.1	18
Philippines	54.8	5.5	9.8	13.5	16.3
Poland	76.5	2.7	4.3	4.2	11.1
Qatar	52.8	2.6	17.2	8.3	18.8
Romania	52	7	13.7	17.3	8.7
Russia	55.7	3.7	9.9	5.3	22.7
Rwanda	60.4	14.2	19.3	3.3	2.8
Singapore	53.5	11.1	15.3	7.9	12.1
Slovenia	68.4	3.3	4.3	4.3	17.9
South Africa	57.4	17.3	12.7	8.8	3.8
Zimbabwe	57.4	8.1	20.6	11.1	2.8
Spain	71.6	8.4	6.1	8.7	5.2
Sweden	55.1	9	7.9	6.7	20.3
Thailand	46.8	11.8	11.9	12.4	16.9
Trinidad and Tobago	59.2	7.7	10.1	11.5	10.9
Tunisia	84.6	1.9	3.2	5.6	4.1
Turkey	51.6	12	7.6	22.8	4.1
Ukraine	63.4	3.3	12.1	2.3	19
Egypt	81.5	2.2	5.9	8.1	2.2
United States	53.1	4	13	18.9	9.9
Uruguay	56.9	12.7	5.2	14.7	9.4
Uzbekistan	40.6	4.4	25	5.7	21.8
Yemen	74.3	1.8	5.7	13.4	3.6

Source: Inglehart et al. (2014). World Values Survey Database

technologies with a lower carbon footprint, and yet many people and organizations continue to depend on in person meetings that require air travel, including meetings to discuss how to address climate change.

It is the way most people respond to those tradeoffs that is of greatest consequence to how we relate to the environment, not just the simple ignorance of the facts about the consequences of our actions or knowledge of our alternatives. Existing evidence suggests we are still very far from living in a world in which most people are prepared to value environmental sustainability over other desired goals, such as high levels of consumption, or jobs. One of the questions in the World Values Survey asks respondents to choose between the statement that ‘Protecting the environment should be given priority, even if it causes some loss of jobs’ and ‘Economic growth and creating jobs should be the top priority, even if the environment suffers to some extent’. For all the countries participating in the survey, on average only 47% of the respondents favored the environment over jobs. The percentage who responded this way varied across countries as seen in Table 1.2, from a high of 74% in Malaysia, to a low of 4% in Haiti.

Table 1.2 Percentage of the population who thinks that protecting the environment should be given priority even if it causes some loss of jobs between 2010 and 2014

Country	Percentage
Malaysia	73.6
Colombia	67
Chile	66.7
Philippines	64.9
Uruguay	64.2
Qatar	63.1
Sweden	62.9
Peru	62.9
Mexico	62.8
Uzbekistan	62.1
Ecuador	61.2
Taiwan	60.6
Brazil	60.3
Kyrgyzstan	59.3
Georgia	59.1
Australia	59
Hong Kong	58.6
India	58.4
Thailand	57.5
China	56.6
Belarus	56.2
Libya	54.5
Argentina	54.2
Kazakhstan	53.9
Morocco	53.2
Ghana	50.3
Russia	50.2
South Korea	48.2
Turkey	48
Ukraine	47.7
Germany	47.7
Estonia	47.7
Cyprus	47
Palestine	46.8
Pakistan	45.7
Slovenia	44.5
Iraq	43
New Zealand	42.6
Singapore	41.1
Netherlands	40.9

(continued)

Table 1.2 (continued)

Country	Percentage
Armenia	40.1
Lebanon	39.9
South Africa	38.3
Poland	37.6
Zimbabwe	37.3
United States	37.2
Jordan	35.8
Spain	35.2
Romania	34.8
Nigeria	33.8
Yemen	33
Tunisia	32
Algeria	31.2
Azerbaijan	31.1
Egypt	30.5
Kuwait	27.2
Japan	22.7
Rwanda	22.1
Haiti	3.8

Source: World Values Survey database. Inglehart et al. (2014)

The relative value people assign to environmental protection over jobs and growth is related to the economic opportunities available, as demonstrated by the case of the United States. Each year, since 1985, the Gallup organization has asked representative samples of Americans whether the environment should be protected, when this goal conflicts with economic growth. During the thirty-four year period over which these data have been collected, a greater priority was given to economic growth during the years when unemployment rates were higher. The figures reported in the previous table from the World Values Survey for the United Sates correspond to a peak in unemployment, over 8% between 2010 and 2013. In 2019, a period of low unemployment at 4%, most respondents, about three in five, agree that environmental protection should be prioritized over economic growth. These fluctuations notwithstanding, however, the percentage of the population who thinks the environment should be protected is very similar in 2019 (65%) to levels in 1985 (61%), even though education and public awareness of climate change have increased during this period (Saad 2019). It is also unclear whether the level of education greatly impacts those choices, though it presumably impacts knowledge of climate change. In the United States, college graduates are only slightly more likely to favor protecting the environment (67%) than those without college degrees (62%).

It is possible that the reason there are few differences in how people with different levels of education, and presumably knowledge, value the environment, is because values, rather than knowledge, play an important role. There are significant differences by age and political affiliation in how likely people are to favor the environment, which suggests that these are valued based choices, not just knowledge-based choices. Among those aged 18–24, 78% would favor environmental protection when it conflicts with economic growth, compared to 58% among those aged 25–54 or to 60% for those over 55. Among Republicans, 35% would favor environmental protection over economic growth, compared to 71% among independents and 82% among democrats. This suggests that if education is to influence how we relate to the environment, it must activate our moral imagination, our capacity for ethical and critical thinking and not just dispense us with more facts.

In addition to the limited predictive value of knowledge and awareness over individual behavior, addressing climate change effectively requires more than influencing the private choices of individuals. Climate change is the result of systems, of production and consumption, and influencing systems requires not just understanding them, but the skills to work with others in inducing change in those systems, not just changes on individual behavior.

1.3 Climate Change Education

A great variety of approaches and methods co-exist under the same term of 'Climate Change Education'. A surface understanding of the term involves teaching students to understand an existing, scientifically established phenomenon, the result of the way in which humans relate to the environment. Indeed, many of the practices in the field represent didactic approaches of teaching facts of that sort. A rich example of that didactic approach is a complete Climate Change Curriculum developed as a collaboration between climate scientists and teacher education faculty at Stanford University, and middle and high school teachers in the San Francisco Bay Area. The curriculum teaches climate science, the impacts of climate change on society and on global resources and mitigation and adaptation strategies (Stanford University 2020).

This approach to climate science education, as teaching the science of climate change, is rooted in the origins of environmental education, a field which emerged in the 1960s. The seminal contributions of Albert Baez, the physicist who initiated UNESCO's efforts to promote science education in the developing world, are illustrative of the interest of scientists on environmental issues at the time. Baez proposed in the 1960s that science education had to help students advance the purposes of Peace, Poverty Reduction, Pollution Reduction and Over Population (Reimers 2007).

This approach of climate science education, and environmental education more generally, as anchored in science education dominates the field to date, as shown by a review of 220 studies of climate change education conducted between 1993 and 2014 which identified that most of them approached climate change as STEM education (science, technology, engineering and mathematics), or with environmental and sustainability education (Rousell and Cutter-Mackenzie-Knowles 2020, p. 198).

This early work to include environmental themes within scientific literacy was stimulated by and reinforced by attention to climate and sustainability by other UN agencies. In 1972, the United Nations convened the first Conference on the Human Environment in Stockholm, which would become a milestone in the global environmental movement. A review of the history of research in environmental education concludes that the UN efforts on environmental issues supported the interest in environmental education (Gough 2013). The same review argues that, over time, UN agencies shifted the field from environmental education, to education for sustainable development, with more emphasis on cultivating human capacities to address environmental and development challenges.

In 2010, for example, UNESCO launched the Climate Change Education for Sustainable Development program as an effort to foster 'climate literacy' among students (UNESCO 2010). The initiative called for four integrated programs: (1) Climate science and knowledge, (2) Climate change education, (3) Climate Change, Cultural and Biological Diversity, and Cultural Heritage, and (4) Climate change, ethics, social and human sciences dimensions.

As the field of climate science education, and education for sustainability, evolved, so did interest on a broader range of outcomes beyond knowledge and on student centered, participatory, cross disciplinary and multidimensional approaches. Illustrative of those are 'whole school' approaches, these are the product of the 'green school' movement, which traces its roots to the 1992 UN Conference on the Environment and Development. These green schools or eco-schools are integrated in networks which form a movement aimed at influencing sustainability in the societies where they operate (Gough et al. 2020, in press). The latter is the approach advocated by UNESCO in a guide to support climate change education:

> "In a whole-school approach, students' classroom learning about climate change is reinforced by the formal and informal messages promoted by the school's values and actions. In other words, students – girls and boys alike - and other members of the school community live what they learn, and learn what they live. The whole-school approach to climate change means that an educational institution includes action for reducing climate change in every aspect of school life. This includes school governance, teaching content and methodology, campus and facilities management as well as cooperation with partners and the broader communities." (Gibb 2016, p. 3).

UNESCO's guide for whole school climate change education emphasizes a six-steps process, and is relatively thin on curriculum and content. The process comprises creating a school climate action team, infusing sustainable development across all subjects, teaching creative and futures thinking, empowering students to take action, address facilities and operations and build community partnerships. In terms of curriculum, the guide offers two examples of activities which could be integrated into each of eleven different subjects, such as designing and maintaining a school garden and compost, creating maps showing areas of the world most at risk due to climate change and examining how societies throughout history have resolved conflicts and responded to environmental challenges (Gibb, 2016 p. 12).

UNESCOs Climate change education program proposed using

> "innovative educational approaches to help a broad audience (with particular focus on youth), understand, address, mitigate, and adapt to the impacts of climate change, encourage the changes in attitudes and behaviours needed to put our world on a more sustainable

> development path, and build a new generation of climate change-aware citizens." (UNESCO 2010, p. 4).

As part of their advocacy for Climate Change Education, and Education for Environmental Sustainability more broadly, the UN and its specialized agencies routinely ask governments to report on the extent to which these topics are included in policies and in the curriculum of instruction in countries around the world. The reports provided by governments to international development agencies suggest that these themes are increasingly recognized by policy and included in the curriculum, which provides an authorizing environment to support instructional practice, but does not necessarily produce high quality or effective practice as will be discussed later in this chapter.

For example, a recent survey administered by UNESCO to member states assessing the extent to which the national curricula in 82 nations (representing about half of the countries surveyed) addressed environmental topics revealed that most governments report that topics such as Climate change, Environmental sustainability, caring for the planet, sustainable development, consumption, and livelihood are included in the national curricula (UNESCO 2018, Figure 6).

In China, Education for Sustainable Development was incorporated in the national standards in 2010. Denmark included climate change education and education for sustainable development in the curriculum in 2009. Recent legislation in Italy mandates the introduction of a required course on climate change (Berger 2019).

> "In India, for example, environmental education was mandated by the Supreme Court in 1991, and in 2003 the government directed the National Council of Educational Research and Training to produce extensive content on environmental education." (UNESCO 2016, p. 25).

In the United States, the new science standards, a set of standards for voluntary adoption by States developed by the National Research Council, the National Science Foundation, the American Association for the Advancement of Science and the National Science Teacher Association, have introduced the subject of climate change in elementary school, with opportunities for deeper study in middle and high school (Chen 2017).

While it is significant that the environmental education movement has gained traction on government policy statements, governments' self-reports to UN agencies following up on inter-governmental agreements have inherent limitations. In addition, governmental commitments to climate change education are fluid, the subject of the partisan contestation that surrounds the topic of climate change itself.

Illustrative of the volatility in the policy priority afforded climate change education is the case of Australia. Over the last two decades, Australia has advanced a number of initiatives on education for sustainability, adopting a national plan of environmental education in the year 2000, focusing more intentionally on environmental sustainability in 2006, and in 2008 increasing the focus on climate change in the Melbourne Declaration: a statement on educational goal for young Australians subscribed by all education ministers (Ministerial Council on Education 2008). The topic, however, remains contentious and many of these efforts were discontinued in

2013. A 2019 report on the state of climate change education in Australia states that the country lacks a coherent strategy for climate change education and that teachers are left to fend for themselves in addressing the subject (Whitehouse 2019). A recent analysis of the state of climate change education in Australia, notes that the references to climate change and integrating sustainability across the curriculum which the Melbourne Declaration of 2008 included, were replaced by another Ministerial statement of goals (the Alice Springs Education Declaration) which removed those references (Gough 2020).

Such fluidity in national support for climate change education notwithstanding, attention to environmental sustainability and climate change education by governments has been stimulated and supported by the growing interest of UN agencies on environmental sustainability and, more recently, on climate change in particular. The landmark 1992 United Nations Conference on Environment and Development, mentioned earlier, advanced a view of development inclusive of environmental sustainability that marked a turning point in the global environmental movement. *Agenda 21*, the action plan resulting from the conference included efforts in education to increase awareness of the concept of sustainable development. At the Conference, the United Nations Framework Convention on Climate Change was presented for signature, entering into force in 1994 once enough countries had ratified it. Article 6 of the convention underscores the importance of education and training to address climate change:

> "EDUCATION, TRAINING AND PUBLIC AWARENESS
>
> In carrying out their commitments under Article 4, paragraph 1 (i) the Parties shall:
>
> (a) Promote and facilitate at the national and, as appropriate, subregional and regional levels, and in accordance with national laws and regulations, and within their respective capacities:
>
> (i) the development and implementation of educational and public awareness programmes on climate change and its effects;
> (ii) public access to information on climate change and its effects;
> (iii) public participation in addressing climate change and its effects and developing adequate responses; and
> (iv) training of scientific, technical and managerial personnel;
>
> (b) Cooperate in and promote, at the international level, and, where appropriate, using existing bodies:
>
> (i) the development and exchange of educational and public awareness material on climate change and its effects; and
> (ii) the development and implementation of education and training programmes, including the strengthening of national institutions and the exchange or secondment of personnel to train experts in this field, in particular for developing countries." (United Nations 1992, page 5).

A decade later, in 2002, the World Summit on Sustainable Development underscored the interdependence of various dimensions of development as part of the notion of sustainability. A full ten years later, in 2012, the UN Conference on Sustainable Development again underscored the interdependence of social, environmental and economic development and highlighted the lack of progress in integrating these three pillars (UNESCO 2016, p. 5).

In 2014, The Ministers and Heads of Delegation attending the twentieth session of the Conference of the Parties and the tenth session of the Conference of the Parties serving as the meeting of the Parties to the Kyoto Protocol, adopted a declaration that specifically underscored the necessity of determined educational initiatives to address climate change. The Lima declaration states:

> "1. Stress that education, training, public awareness, public participation and public access to information play a fundamental role for all countries to achieve climate-resilient sustainable development and contribute to meeting the objective of the Convention;
>
> 2. Reaffirm our commitment to promote and facilitate at the national, subregional and regional levels the development and implementation of educational and public awareness programmes on climate change and its effects, of public access to information and of participation in decision-making on climate change;
>
> 3. Encourage all governments to include the issue of climate change in curricula and to include awareness-raising on climate change in the design and implementation of national development and climate strategies and policies;
>
> 4. Urge all Parties to give increased attention to the topic of education, awareness raising and public participation in all aspects of climate change negotiations;
>
> 5. Call on all Parties to re-emphasize the importance of education, training, public awareness, public participation and public access to information on climate change and its effects in the new global agreement to be concluded in Paris in 2015;
>
> 6. Reaffirm our commitment to cooperate and engage through bilateral and regional complementary initiatives that aim to raise awareness and enhance education on climate change and its effects." (United Nations, Conference of the Parties 2014).

In September of 2015, at the annual general conference of the United Nations, the governments of the nations participating adopted a compact of development which embraced the goal of sustainable development, identifying seventeen goals and a series of specific targets. Sustainability was the driving concept of the entire framework, which highlighted the role of education, one of the sustainable development goals, in achieving all of the remaining goals, including Goal 13 'Take urgent action to combat climate change and its impacts'. Specifically, Goal 4 'Ensure inclusive and equitable quality education and promote lifelong learning opportunities for all', includes a target (4.7) that explicitly focuses on education about sustainable lifestyles:

> "By 2030, ensure that all learners acquire the knowledge and skills needed to promote sustainable development, including, among others, through education for sustainable development and sustainable lifestyles, human rights, gender equality, promotion of a culture of peace and non-violence, global citizenship and appreciation of cultural diversity and of culture's contribution to sustainable development." (UN 2020, Target 4.7.)

That same year, the United Nations Climate Change Conference in Paris approved the Climate Paris agreement, eventually adopted by delegates from 195 nations recognizing the importance of education for climate change education in Article 12.

At present UNESCO promotes Climate Change Education through its Education for Sustainable Development Program, established in 2010, through policy advocacy, facilitating exchange of good practice on climate change, supporting countries through capacity building, supporting the Associated Schools Network in climate action, and disseminating education resources. Salient among those resources is a guidebook to support whole-school approach to ESD and climate change, which involves mainstreaming sustainability into all activities of the school, including curriculum and teaching, advancing sustainability in facility management, school governance and cooperation with parents and community (Gibb 2016).

An analysis of the government reports submitted as part of the process of reporting under the UN Framework Convention on Climate Change, shows that climate change education is addressed by most countries, 95% of those reporting, largely through public awareness efforts that emphasize cognitive dimensions. There is, however, relatively limited emphasis on socio-emotional and behavioral dimensions. Among the 194 countries reporting to the UN Framework Convention on Climate Change, 95% of them report incorporating climate change education in one of their recent reports (UNESCO 2019a, p. 5). Half of those did incorporate climate change education in formal education settings. Considerably less attention was given to education of groups in government, industry, non-governmental organizations and the scientific community (UNESCO 2019a, p. 5). Those reports prioritize public awareness (47%) followed by education (17%) and training (15%). The reports on inclusion of climate change in formal education prioritized knowledge, followed by skills, and to a much lesser extent by socio-emotional skills. At the primary level, for instance, 67% of the reports were about knowledge, followed by 27% focused on skills and 7% on socio-emotional skills. At the secondary level, 63% focused on knowledge, 33% on skills and 4% on socio-emotional skills. At the tertiary level, 75% focused on knowledge, 0% on skills and 25% on socio-emotional skills (UNESCO 2019a, p. 7).

An in-depth analysis of policy documents in ten countries with an expressed commitment to Education for Sustainable Development and Global Citizenship Education undertaken by UNESCO, revealed that in all these countries there are abundant references to both of these concepts, and that they are expressed in terms of cognitive, socio-emotional and behavioral dimensions (UNESCO 2019b). In the documents examined in these countries—Costa Rica, Japan, Kenya, Lebanon, Mexico, Morocco, Portugal, Republic of Korea, Rwanda and Sweden—there were almost twice as many references to Global Citizenship Education (representing about 60% of the references) than to Education for Sustainable Development (representing about 30%) across national laws, strategic plans and policies, national curriculum frameworks, programmatic documents and subject specific curriculum. These references were present across various subjects in the curriculum, and the emphasis on cognitive dimensions, relative to socio-emotional and behavioral dimensions, increased in secondary education.

Growing attention to climate change education on government policy frameworks and curriculum has created space for the development of a practice of climate change education. However, this practice is very heterogenous and evidence on its effectiveness is contested. As a result, the field has not yet reached the point where there is a robust consensus on what the evidence indicates is good practice. This absence of a robust academic consensus contributes to the fragility of this field of educational practice, to disconnects between policy and practice, and to the vulnerability of the field to the shifts induced by partisan influence in education policy.

A recent review of environmental education research, for example, concludes that the focus of most empirical studies is on individual effects in energy conservation behavior among children and youth, with very limited attention to effects on collective action or on the kind of sociotechnical transformation necessary to move away from fossil fuel energy consumption to renewable based energy systems (Jorgenson et al. 2019). The reviewers argue that many of the existing Environmental Education approaches depend on dated approaches developed in the 1970s and 1980s which assumed that "environmental problems could be adequately addressed through resource conservation and incremental changes to technology and human behavior." (Jorgenson et al. 2019, p. 160) The reviewers argue that this exclusive focus on individual behavior is inadequate to address climate change, "a systemic problem of such scale and complexity that addressing it requires systems level change that results from the interaction and coordination of actions and innovations across multiple levels of scale" (Jorgenson et al. 2019, p. 160). Of the 70 studies included in the review, which covered articles published between 2012 and 2018, less than a handful conceived of environmental/climate change education as influencing collective action.

By minimizing the role of collective action, environmental educators and researchers may be reinforcing a simplistic and narrow conception of the relationship between climate change, human action, and energy system change and distorting the fact that many of the most impactful climate actions are decisions about energy supply systems that are made by state and market sector actors under direct pressure from advocacy coalitions and other social collectives (Jorgenson et al. 2019, p. 166). Another review of the climate change education literature suggests the need for approaches which are socially transformative, focusing on empowering students to act. The review suggests that much climate change education is focused on helping students understand climate change, the science of climate change, but is insufficiently focused on helping them identify pathways to change climate change (Stevenson et al. 2017, p. 70).

Another recent review of climate change education studies, covering literature published between 1993 and 2014, concludes that many of these studies document very limited effects of the programs evaluated on students' attitudes and behavior (Rousell and Cutter-Mackenzie-Knowles 2020). The authors of the review argue that largely missing from the literature are approaches to climate change education which are participatory, interdisciplinary, focused on affect and creative. This review of the literature on climate change education identified a tension between the more predominant knowledge-based approaches to science education and interdisciplinary, affect driven and experiential education. Among the former, focused on the development of scientific knowledge about climate science, some studies however found no relationship between scientific knowledge and pro-environmental behavior. In contrast, a number of the studies focusing on cooperative, interdisciplinary, placed-based, experiential programs showed that those impacted attitudes and behaviors towards climate change (Rousell and Cutter-Mackenzie-Knowles 2020, p. 196).

About half of the studies reviewed focused on fostering scientific knowledge-based instruction on climate change education, followed by an emphasis on curriculum and pedagogy. These two approaches are followed by behavior change approaches, emphasizing education approaches designed to influence individual behavior, and adaptation and mitigation approaches emphasizing minimizing the impact of climate change.

The approaches to climate change education which focus exclusively on the development of knowledge are based on limited models about the factors which influence engagement with climate action. Emotions have been found to be important correlates of active engagement with climate. A study of the relationship of hope concerning climate change with pro-environmental behavior among Swedish youth and young adults found that hope plus worry about climate change was positively related to pro-environmental behavior, whereas hope plus lack of worry was not (Ojala 2012). Given the important role of emotions such as hope and worry in pro-environmental action, Ojala has advocated that they should be cultivated to sustain actions that challenge existing norms and institutions that contribute to climate change. She explains that a transformative and transgressive climate change education that recognizes the worry and anxiety that disrupting and transgressing generates, can create hope through critical emotional awareness and through activities that develop visions of preferable futures (Ojala 2016, p 52).

While the field of climate change education lacks the scientific consensus which characterizes domains such as 'literacy' or 'how people learn' (reflected, for instance, in the consensus reports on these topics produced by the National Research Council in the United States) there are emerging efforts to synthesize ideas about what works in practice. For example, a recent review of literature on the practice of climate change education, conducted by the Alberta Council for Environmental Education (2017) identifies six key principles of excellent climate change education:

(a) Frame climate change education in ways that focus on solutions, rather than on problems, build a positive narrative around shared identity. Focus on energy, conservation and outdoors education. Rely on pedagogies which engage in deliberative discussions, promote exchanges with scientists, address misconceptions, and implement school and community projects
(b) Keep the audience in mind. Develop curriculum that is appropriate to the age of the child, support teachers.
(c) Design programs which are action oriented. Build the agency of students.
(d) Develop activities that extend beyond climate science, including imagining a positive desired future, focus on local content, teach students how to think, not what to think, do not scare students.
(e) Establish connections to the curriculum and identify competencies. Emphasize cross-curricular approaches, cultivate systems thinking, and help students understand the interdependencies between climate change mitigation, adaptation and resilience.
(f) Evaluate for program improvement.

In spite of this growing body of practice, and of the more limited body of research, documented levels of student knowledge and skills with respect to Climate Change or Environmentally Sustainable Education more generally appears to be inadequate to meet the urgency of the challenge, and as mentioned when discussing the reviews of this research, evidence of impact of climate change on attitudes or behavior is elusive.

The Program for International Student Assessment, administered by the OECD, shows that, on average, only one in five students in the OECD countries can consistently identify, explain and apply scientific concepts related to environmental topics (OECD 2012). Conversely, 16% of the students don't have enough knowledge to answer questions containing scientific information related to basic environmental issues, and 20% of the students are just at that baseline level of scientific proficiency. These low levels of scientific knowledge and skills are in spite of the fact that all students in the OECD attend schools that report that they teach environmental science as part of the science curriculum. The latest administration of the PISA study revealed that less than 10% of all students tested could distinguish facts from opinions (OECD 2019, p. 3).

This evidence underscores the fact that the current challenge with climate change education is not just a challenge of including it in the curriculum, and most certainly not a challenge that is solved when governments include it in policy pronouncements, but it is also a challenge of developing the capacity of teachers to support deeper learning among their students. Deeper learning requires an integrated view of how knowledge relates to behavior. Such integrated views about the breath of competencies that undergird human functioning, extending them beyond knowledge, have evolved over the last few decades.

In 1991, UNESCO's General Conference, proposed the creation of a Commission to develop a framework for education in the 21st century. The report produced by the commission proposed that education should be organized around four goals: learning to know; learning to do; learning to live together; and learning to be (Delors 1996). Building on UNESCO's Delors Report, UNICEF developed a framework of life skills and citizenship to support the development of children in the Middle East that reflects an ambitious set of twelve core life skills aligned to the four pillars in UNESCO's report. Learning to know, for instance, is reflected in Skills for Learning (creativity, critical thinking, problem-solving), learning to do in Skills for Employability (cooperation, negotiation, decision-making), learning to be in Skills for Personal Empowerment (self-management, resilience, communication) and learning to live together in Skills for Active Citizenship (respect for diversity, empathy, participation) (UNICEF 2017, p. 4).

The Organization for Economic Cooperation and Development also contributed to the global dialogue on broader goals for education through a Learning Framework outlining an expanded set of competencies that could contribute to individual and collective wellbeing (OECD 2020).

> "The OECD Learning Compass 2030 defines core foundations as the fundamental conditions and core skills, knowledge, attitudes and values that are prerequisites for further learning across the entire curriculum. The core foundations provide a basis for developing student agency and transformative competencies. They are also the building blocks upon which context-specific competencies for 2030, such as financial literacy, global competency or media literacy, can be developed." (OECD 2020, p. 2)

In 2019, UNESCO established a high-level commission to prepare a report on the Futures of Education. In a series of submissions to the commission by several holders of UNESCO's chairs, many of them highlight the importance of educating for climate-change and advance ideas that emphasize going well beyond knowledge of the facts about climate change. Some of these authors argue that sustainability is an inherently cross-disciplinary topic, which requires an understanding of the systems which undergird climate change. This requires being able to integrate insights from economics, science and social science, but traditional curricular silos impede such understanding of systems (Jain 2020, p. 30). Such interdisciplinarity is necessary not just to cultivate understanding of systems among students, but also among scientists themselves. Given the importance of geology to understand sustainability, it is paradoxical that most geologists are not involved in sustainability science and their education seldom addresses sustainable development (Stewart 2020, pp. 39–40). In order to mobilize geology based knowledge to address human harm to the environment, requires collaboration with engineers, planners, biologists, zoologists, ecologists, agronomists, environmental scientists as well as social and behavioral sciences (Stewart 2020, p. 40). Educating for sustainability requires also cultivating "diversified and shareable imaginations of the territory and our living environments that contribute to the development of a viable future" (Poullaouec-Gonidec 2020, p. 33). The development of such imaginations of sustainable living environments requires dialogues that are anchored on the arts, sciences and humanities. Central to educating for sustainability is also to help students develop ethical frameworks which can value the environment and life in themselves, an education which helps students know the multiple humanistic traditions and that engages them with environmental ethics (Mantatov et al. 2020, p. 35). Educating to value the natural world, instead of material possessions, is also foundational to the development of ethics and imaginations that value nature and the environment.

The breadth of skills identified in these various frameworks underscores the importance of addressing more than knowledge of facts about climate change in order to prepare students to translate knowledge into action. The capacity to engage students in such deeper learning requires high level of skills from teachers, and a well developed curriculum, in order to not only engage their students in high cognitive activation tasks, but also to develop the socio-emotional competencies that undergird agency and the capacity to take responsibility for climate change and to collaborate with others productively in addressing it. The reviews of research on climate change education cited earlier show that such approaches to climate change education, while apparently effective, are rare.

In order to engage their students in deeper learning in this domain teachers need support to develop their own knowledge about climate change. A recent survey of a nationally representative sample of science teachers in the United States conducted by the National Center for Science Education (Plutzer et al. 2016) revealed that while three quarters of the science teachers address climate change in their classes, only half of them do so in ways that are aligned with the current scientific consensus. When asked to rate their own content knowledge with respect to climate change, ecology, modern genetics, weather forecasting and health and nutrition, 17% of the teachers report that they know less about this topic than most other high school teachers, and 31% report the same for weather forecasting models. Only 28% of the teachers report that their knowledge of climate change is very good or exceptional, compared to 45% who report this level of knowledge for ecology or 44% for genetics or 48% for health and nutrition (Plutzer et al. 2016, p. 19). When asked to select a series of possible topics to be covered to teach a unit on greenhouse gases and recent global warming, a topic which most teachers reported they taught and one on which the basic science on how these gases trap heat is a century old and noncontroversial, only some of the teachers selected as high priority topics which are essential to understand greenhouse gases. (Table 1.3)

The same survey reveals that most teachers are unaware of the scientific consensus attributing global warming to human activities, with 61% of them demonstrating ignorance of such scientific consensus. Only 39% of the teachers in the study correctly recognize that over 80% of climate scientists think that global warming is caused mostly by human activities, and an additional 21% of the teachers admit that they don't know the answer, with the remaining 40% providing an incorrect answer (Plutzer et al. 2016, p. 22). Teachers report that they have received very limited training on climate change, only 43% had any formal instruction on the subject at the college level, and only 10% completed a course on the subject (Plutzer et al.

Table 1.3 Priority given to potential topics in a teaching unit on the greenhouse effect. (Plutzer et al. 2016, p. 21)

"Imagine that you were asked to teach a 2–3 day unit on greenhouse gases and recent global warming. What priority would you give to including each of the following possible topics?"					
	A high priority	A medium priority	It is not necessary to cover this topic	This topic should not be covered	I do not have an opinion
Carbon dioxide trapping heat in the atmosphere	74%	22%	1%	0%	3%
Use of coal and oil by utility and electric companies	59	40	2	0	3
Emissions from industry	56	33	2	0	4
Destruction of forests	55	39	2	1	3
Depletion of ozone in the upper atmosphere *(foil)*	42	41	11	3	3
Incoming shortwave and outgoing longwave energy	24	39	15	1	20
Use of chemicals to destroy insect pests *(foil)*	23	42	23	5	?
People heating and cooling their homes	21	62	10	1	6
Use of aerosol spray cans *(foil)*	14	56	20	4	7
The impact of launching rockets into space *(foil)*	4	27	41	7	22

2016, p. 23). Among those without education on climate change during initial preparation, only 18% received any professional development on the subject. Teachers recognize this topic as a high need for preparation, and 67% report that they would be interested in professional development opportunities on the subject (Plutzer et al. 2016, p. 24).

1.4 The Limitations of Current Climate Change Education Efforts

In addition to the limitations stemming from too narrow a definition of the intended outcomes of climate change education (focused only on low levels of cognition with insufficient attention to higher order cognitive skills or to intra personal or inter personal skills) and of the approaches to achieve them (didactic and siloed into a single subject instead of problem and activity based and interdisciplinary), a second set of limitations explains the challenges of going to scale with effective climate change education programs.

As described in the previous section, many of the ongoing climate change education programs are based on an implicit or explicit top down model of change which assumes that if intergovernmental bodies and governments embrace the purpose of climate change education, this will transform instruction and learning. These efforts are misguided in two ways. First, they assume that climate change education is a technical challenge with a universal solution, that there is one content and modality of education which can be rolled out across all jurisdictions in the world. Technical problems require that the solution to the problems is known. But if climate change education were a technical challenge, and we knew how to solve it, we would not face the paradox described in the first section of this chapter given how much policy rhetoric has been devoted to it and how much attention governments profess to devote to it in their reports to UN agencies and in policy declarations. The simple question we must answer is, given that governments state their interest in climate change education, and given that there is an abundance of resources that offer guidance on how to do it, such as UNESCO's guide on a whole school approach to climate change education, why isn't it happening and why isn't it achieving the necessary results? Why is it that "Despite efforts over the past 40 plus years, environmental sustainability is still on the margins of the curriculum in most countries." (Gough 2016, p. 84).

International declarations, governmental agreements, and top down policy guidance on climate change education suffer from an inherent limitation. Effective programs of climate change education need to be designed to serve specific populations and to fit the particularities of institutional settings in unique jurisdictions. The need for contextually relevant approaches to climate change education stems from three facts. One, that climate change influences various locations and populations in unique ways. Two, that the systems that contribute to or mediate the impact of climate change are local, and changing them requires knowledge and skills that are fit for context and purpose. Third, that the approaches to climate change education need to fit the characteristics of the local education systems or schools.

I learned about these limitations of designing 'generic' curricula from my own work over a decade designing curriculum aligned with the United Nations Universal Declaration of Human Rights and Sustainable Development Goals. In 2009–2010, with a group of my graduate students, I developed a comprehensive curriculum, spanning from kindergarten to high school, aligned with the UN SDGs (we initially worked with the Millennium Development Goals, and later on substituted them with the Sustainable Development Goals as they were adopted at the UN General Assembly in 2015), with the Universal Declaration of Human Rights, and with the World Economic Forum Risk Assessment Framework. From the study of those goals, we developed a framework of competencies which a high school graduate should have in order to contribute to achieving such goals, we emphasized cognitive and socio-emotional competencies. Then, we used this framework to guide the development of 350 units to be taught in a special course, a 'world course', that would provide students explicit opportunities to integrate knowledge gained in various disciplines, as they worked on projects aligned with those competencies (Reimers et al. 2016). The book was well received and as a result many educators attempted to incorporate this curriculum in their schools. I discovered then that to do so they had to make significant adaptations to our original design, for instance, most could not devote the eight to ten hours a week, every week from kindergarten to high school, we had anticipated would be necessary to teach this curriculum. Most teachers ended just using this comprehensive curriculum as inspiration and as a resource, and adapted it in ways that fit with their existing goals and capacities. Even the first school that was genuinely motivated to adopt the 'world course', a newly established network of international schools (the Avenues School) ended up making successive adaptations to the original design.

Thinking that the challenge the 'world course' had encountered was one of complexity, in 2016, working with 36 of my graduate students, I developed a streamlined global education curriculum, from kindergarten to high school, following the same process of backward design from the UN Sustainable Development Goals (Reimers et al. 2017). This second attempt worked much better, in part because this time the book included not just a curriculum prototype, but a process of whole school change proposing how to introduce the new curriculum in the school and how to integrate it into the existing curriculum. One significant shift represented by this new curriculum is that I moved away from a dedicated 'world course' that required its unique place in the curriculum, dedicated teachers and eight to ten hours a week, every week from kindergarten to high school, replacing it with an approach that proposed to infuse global education widely across the curriculum. We offered a method to do this, and five lessons per grade which could be infused into existing subjects. This shift reduced some of the institutional demands to implement global education in the school, as it required no additional time and no additional faculty, even if it engaged considerably more people in the process, in effect increasing the demands. The approach differed also in that it made it possible to develop a whole school approach to sustainability education that did not see schools as blank slates, but that allowed schools to build more intentional sustainability education programming on already existing missions, plans, courses and activities. In an effort to convey that the process of whole school design of their own approach to sustainability education was more important than following any specific curriculum, I then developed, with a group of 34 of my graduate students, a variety of different curriculum prototypes, aligned with the UN Sustainable Development Goals (Reimers et al. 2018). My goal with this third curriculum resource was to convey that there were multiple pathways that schools could follow to empower their students as global citizens, that would be upstanders for sustainability.

Again I found that the most productive use of these resources was when teachers adapted it to their own needs, goals and capacities. Based on that work I then developed approaches to help teams of teachers, across schools, collaborate in developing their own adaptations of those curriculum resources, make them their own, and improve them through their practice.

From this work I discovered how important it was to attend to the details of implementing a new curriculum in a particular context. Using my book 'Empowering Students to Improve the World in Sixty Lessons' as a starting point, I have worked with networks of teachers in developing global education curriculum, such as the Rete Dialogue, a network of teachers in Italy committed to democratic education, in translating and adapting this book to the Italian context. Over a year, this network of teachers translated the original book, taught these lessons, and then modified them, as part of a learning community in which they collaborated in this process across various regions in the country. The result of this process was a revised curriculum, reflecting the learning these teachers had drawn from their practice in experimenting with the original lessons (Reimers et al. 2018).

Similarly, working with a group of fifty teacher leaders supported by the National Education Association Foundation in the United States, we developed a curriculum, inspired by 'Empowering Students to Improve the World in Sixty Lessons' in which teams of teachers from all US states collaboratively designed grade specific lessons aligned with the UN Sustainable Development Goals, taught them in their respective schools, and then improved based on their various experiences teaching them. This year long collaborative project, relying on the use of communication technology, led to two publications developed with two different group of teachers which they then used to further advance global education in their schools (Reimers et al. 2018, 2019).

These experiments designing and supporting the adoption of a sustainability curriculum in a variety of schools in different countries taught me that much more than curriculum was necessary to support a change in the culture of education.

Developing an approach to climate change that is fit for context is analogous to designing a new toilet that is fit for particular contexts, to use the example described earlier, rather than to transplant a universal design of a toilet from one setting to another. While there may be general principles of climate change curriculum that can be usefully taught across the world, the contextual nature of the impact of climate change and of mitigation approaches limits the value of universal curricula and approaches.

The second way in which many of the existing efforts to mandate climate change education from on high in national governments are misguided is that this top down approach to climate change education is equivalent to thinking that if we only wished for climate change education to happen, these wishes would trickle down to every school around the world. Trickle down climate change education does not work, even if it is enhanced with a few demonstration schools and with some one size fits all instructional materials.

As with other failed efforts to transform educational practice, the chief shortcomings of ongoing climate change education efforts are in underestimating the capacity and institutional requirements at the point of delivery: the classroom and the school, or the instructional setting of the nonformal education program. For example, UNESCO's guide on using whole school approaches to climate change does not include a single word on how to support teachers to develop the capacities to teach climate change. Similarly, a recent review of 221 studies on climate change finds that the topic of teacher professional development is conspicuously absent, and the few studies that addressed it focused on pre-service teacher preparation rather than on professional support for practicing teachers (Rousell and Cutter-Mackenzie-Knowles 2020, p. 200). A similar conclusion is reached by the authors of a recent study of the place of environmental sustainability in teacher education programs in Canada, who found that over the last four decades, Canadian faculties of education have faced many challenges incorporating environmental sustainability education, and as a result the topic remains a very peripheral concern in pre-service as well as in-service teacher education programs (Karrow et al. 2020, p. 2).

The same conclusion that professional development is an afterthought, or entirely missing, from efforts to advance environmental education is reached by the author of a study of the history of environmental education in Australia and England who found that, even when the curriculum included the goal to make sustainability a priority "the actual content of the four core areas of the Australian Curriculum (English, History, Mathematics, and Science) does not enact the statement's intent, nor is there guidance for teachers in implementing the Organising Ideas for Sustainability" (Gough 2016, p. 89). This study concludes that the ambitious framing of environmental education as a cross-curricular priority in practice was reduced to eight areas of content in the subject of science, and found no specific strategies to support teacher collaboration across subjects. (Gough 2016, p. 91).

Notwithstanding the fact that international agencies and governments seem to assume teacher capacity to teach climate change, the development of effective climate change curriculum appears to exceed the pedagogical capacities of most teachers, schools and education systems, and hence a new climate change curriculum cannot be ruled by decree, but must instead be supported with opportunities for teachers to develop new knowledge and capacities. This is simply a corollary of what is known about how to build instructional capacity for deeper learning and twenty-first century education (Reimers and Chung 2018; Reimers 2020a).

As is true of other ambitious efforts to change the curriculum, such as educating for global citizenship, in order to transform educational practice, the design and implementation of climate change education programs needs to be approached multidimensionally, attending to the cultural, psychological, professional, institutional and political dimensions of the change process (Reimers 2020b, c).

Transforming school culture at scale, for large numbers of schools in education systems, will require more than effective curriculum and professional development. In a recent study of approaches to global citizenship education I argue that past approaches to global education have failed because they have adopted too narrow a view of the process of change, focusing on curriculum, or on teacher professional development, but failing to simultaneously address the change process through the five dimensions which characterize it: cultural, psychological, professional, institutional, and political. These five perspectives integrate what is known about how students learn and how schools change (Reimers 2020b).

> "The cultural perspective, for example, defines the broader set of societal hopes for schools, norms and values which define what are accepted educational goals and practices. The psychological perspective illustrates the theories of learning which undergird the learning and teaching process. The professional perspective focuses on how expertise is inserted in professional roles to advance teaching and learning. The institutional perspective attends to the various structures, processes and resources that provide resiliency to the system of education, governing the interactions among the actors that form the system and providing stability and meaning to teaching and learning. The political perspective illustrates how the interests of various groups are negotiated and conflicts resolved, resulting in a particular culture of education." (Reimers 2020b, page 8).

> "Together, these five perspectives illuminate the complete process of change as the partial elements highlighted by each perspective offers a perspective that complements what other perspectives enlighten and, together, these various elements brought to light by each perspective interacts with the elements highlighted by other perspectives. Paraphrasing Goethe who said that the person who speaks with only one language sees the world with one eye, thinking about educational change through a singular frame is seeing change with one eye. A multidimensional model thus helps capture the gestalt of the process of educational change and provides depth, perspective, a fuller and more complete understanding." (Reimers 2020b, page 22).

A cultural perspective requires that we understand the relationship between a program of climate change education and the expectations and understanding of the local community, and that we engage them in broadening the zone of acceptance of what can be taught so that the curriculum can make room for climate change education. Many students and parents are already aware of the urgency of addressing climate change and, in many settings, they are likely to be allies, supportive of climate change education efforts. A recent survey of a representative sample of parents and teachers in the United States, found that 74% of the teachers, and 68% of the parents, believe that schools should teach about climate change and its impacts in the environment, economy and society. An additional 12% of the teachers and 16% of the parents believe that climate change should be taught, but not its impact. In spite of such support for teaching climate change, only 42% of the teachers and 45% of the parents in the same survey talk to children about climate change (Kamenetz 2019).

A psychological perspective draws on the advances of the learning sciences to design learning experiences that effectively help students gain the deep understanding necessary to inform and motivate behavioral changes. A professional perspective focuses on how to advance climate change education in a way that reflects the best expert knowledge on climate change, and in a way that strengthens the professional capacities of teachers and other educators.

An institutional perspective focuses on the systemic nature of the process of change, requiring alignment between curriculum, assessment, instructional resources, teacher support, and leadership.

A political perspective focuses on identifying the various stakeholder groups affected by climate change education and their position with respect to a climate change education program, and with creating opportunities to mobilize as much support as possible and address the interests opposing it (Reimers 2020b).

When it comes to addressing themes which may challenge prevailing views, or dominant political structures, teachers and schools must tread lightly, as there are boundaries to the educational ideas and goals that communities will find acceptable. This shouldn't mean that teachers falsify the available scientific consensus with respect to climate change or the fact that most scientists agree it is caused by human activity if powerful interest in their community are opposed to teaching such scientific consensus, but it underscores the necessity of developing synergies between school-based efforts to address climate change and non-formal education programs. Communities need to be educated as well in order to expand the zone of acceptance of what schools are authorized to do.

The failure to address educational change through a cultural and political perspective can lead to implementation processes which transforms curriculum and policy in dramatically undesirable ways. For instance, in the United States the politization of discussions of climate change leads teachers to teach content which deviates from the scientific consensus. A recent study of the National Center for Science Education of how teachers teach climate change in the US found that whereas three quarters of the science teachers did address climate change in the curriculum, only 54% did so in ways which were aligned with the scientific consensus. In contrast, 10% of the teachers taught incorrect knowledge, such as the ideas that recent increases in temperature are due to natural causes and teach that it is not the case that the scientific consensus establishes that global warming is primarily being caused by human release of greenhouse gases from fossil fuels. An additional 31% of the teachers sent mixed messages in their teaching, correctly teaching that the scientific consensus is that that recent global warming is primarily being caused by human release of greenhouse gases from fossil fuels, but incorrectly teaching that many scientists believe that recent increases in temperature are likely due to natural causes (Plutzer et al. 2016, p. 16).

The challenge of climate change education thus becomes a challenge of augmenting the capacity of education systems and schools to develop and teach high quality curriculum, evaluate it, and develop and implement high quality programs of teacher education that support teachers in adopting more effective pedagogies of climate change education. This challenge of capacity in classrooms and schools must be addressed in order to scale a new set of instructional practices and a new culture of education that support climate change education. The culture of schools does not change because governments subscribe international conventions, or because they include climate change education in their policy pronouncements or in the curriculum or even because they distribute lesson plans to schools. Transforming the culture of education to advance climate change education requires that we think about the enterprise through a multidimensional framework that attends to the cultural, psychological, professional, institutional and political aspects of the enterprise (Reimers 2020b). Teachers and school leaders need knowledge and skills to effectively deploy programs of climate change education addressing these five dimensions of the process, and this is currently absent in many schools.

1.5 The Need for New Strategies for Climate Change Education

The paradox presented by the fact that in spite of the recognition that education can address climate change our efforts have been insufficient to address the most serious risks posed by human-made environmental degradation, should not lead us to abandon our efforts, or to conclude that education is irrelevant to the challenge of climate change. Rather, it should cause us to rethink our approach to this adaptive challenge. This strategic rethinking of education for climate change needs translate into situated responses focusing on who should be educated, what the focus of such education needs to be, and how such education should be delivered. Because the ways in which climate change impacts people vary in different jurisdictions, and because the characteristics of educational institutions available to educate vary in their reach and capacity, education strategies need to be developed which are specific to specific education contexts. International organizations are better at forging agreements than they are at designing and supporting the implementation of context specific climate change education strategies in particular localities. As we have seen such efforts in policy rhetoric serve a useful, but insufficient, role. They open up space for climate change education, but they do not produce effective climate change education. Context-agnostic guidance is of limited use to effectively support the implementation of changes that transform the capacity of teachers, schools and non-formal education institutions so they can help all gain the knowledge, skills and dispositions to adapt to, mitigate and revert climate change.

Many different populations across a variety of jurisdictions must be educated to address climate change. They include children and youth, certainly, as well as adults in their various occupations and roles so they can from within those roles adopt practices that are more sustainable. In their roles as consumers, farmers, factory workers, commuters, community members, business owners, CEOs or political leaders or activists, each person needs to not only understand climate change, but have the skills to translate that knowledge into implications for their own behavior, and the capacity to make a commitment to acting in more sustainable ways, including not just individual consumption but collaboration with others and collective action to transform systems. Climate change education needs to involve school-based and university-based programs, as well as non-formal and informal education modalities. It needs to educate old and young, the employed and the un-employed, leaders and followers, business owners and workers.

To be sure, there is a need to educate people on the scientific consensus with respect to climate change, as summarized for instance by the National Aeronautics and Space Administration Agency (NASA 2020). But knowledge of facts based on scientific consensus alone is likely to be insufficient as we have seen. People need the kind of deeper knowledge that can help them think and act in more environmentally sustainable ways. Education for climate change must therefore cultivate the broad range of skills which contemporary frameworks posit are necessary to participate in a rapidly changing world.

In order to be more effective at supporting such deeper learning, climate change education requires adequate supports to build the capacities of teachers and schools to advance pedagogies that foster high cognitive activation and that develop life skills to translate such knowledge into changed patterns of behavior. These efforts in schools and universities need to be enhanced with similarly effective non-formal programs which reach most adults. In other words, we need to considerably augment the intensity and efficacy of climate change education efforts at the point of closer proximity with learners around the world, at multiple points of delivery. We will not curb climate change with a few pilot projects, or with small networks of schools committed to the enterprise, or with episodic media campaigns or with pronouncements from international organizations or governments, or with curriculum intentions.

We also need a rigorous assessment of the underlying logic of ongoing efforts, of the underlying program theories of how certain actions are expected to achieve the results we seek, and a shift in mindset that makes visible how we weigh various criteria involved in making individual and collective choices and that give greater weight to sustainability in human behavior.

A context specific strategy needs to address these questions:

1. What are the specific impacts of climate change in this jurisdiction? How do they impact various human populations? Which of these populations needs to be educated on climate change?
2. What knowledge, dispositions and behaviors could mitigate the impact of climate change and are there ways in which changes in the behaviors of populations in this jurisdiction could slow down climate change?
3. What are the means of delivery to reach each of the specific populations in this jurisdiction?
4. What curriculum can best educate each population?
5. What institutions can support the development of the institutional capacity necessary to deliver such curriculum effectively?
6. What institutional collaborations can support the implementation of this strategy?

Developing and implementing such a comprehensive strategy is an adaptive challenge that will require a multi-stakeholder coalition that can produce collective leadership as no single institution has the authority, the power or the capacity to alone address the multiple efforts required. The design of such a strategy requires addressing efforts to advance climate change education attending to the cultural, psychological, professional, institutional and political dimensions of the enterprise (Reimers 2020b). Universities are singularly positioned to work with local governments, and with schools, to augment the capacity necessary to develop and deliver climate change education curriculum.

1.6 The Need for Systemic, Multilevel and Multidimensional Perspectives In Climate Change Education

If climate change education is to support deeper learning among students, it cannot be a simple add on to the curriculum, a new silo where students access new content in some subjects. It certainly cannot be limited to providing students with knowledge of the facts about climate change or even with an understanding of how the systems that impact climate function. The reviews of policy and practice discussed earlier suggest that the earlier efforts of Education for Sustainable Development and Climate Change Education addressed the task primarily as developing new cognitive skills, as providing students with more knowledge. As understanding of curriculum has evolved to embrace a whole-child approach, and to focus on the breadth of skills essential to participate in rapidly changing contexts, Climate Change Education efforts are also embracing a multidimensional view focusing on cognition, socio-emotional and behavioral dimensions.

Climate change education curriculum needs to be aligned with such contemporary frameworks of twenty-first century skills and deeper learning in order to contribute to adaptation, mitigation and reversal of climate change, and teachers need to be supported to develop their skills in engaging their students in deeper learning about climate change.

Effectively creating the opportunities for the development of that breadth of skills with respect to climate change will require the type of whole school approach to change advocated by UNESCO (Gibb 2016). Supporting schools in enacting those changes will require strategies that are responsive to specific contexts, and that include effective professional development for teachers and school leaders. A recent study of programs which successfully prepare teachers to educate the whole child in various countries around the world concludes that they all indeed adopt a whole school approach to educational improvement (Reimers and Chung 2018). In addition, these programs have the following characteristics:

- They reflect a conception of adult learning that sees it as socially situated and responding to current needs of teachers for learning.
- They involve sustained and extensive opportunities for teachers to build capacities, often extending an entire school year, or spanning across multiple school years
- The modalities of professional development are varied. They include independent study of new material, discussion with peers and others, individual or group

coaching, demonstrations of new practices, independent research projects and opportunities for reflection.

- The curriculum of the programs examined covers a blend of capacities, from a broad focus on helping students develop particular capacities to a highly granular identification of particular pedagogies and instructional practices that can help students gain those skills.
- The curriculum of these various programs reflects a view of learning which includes cognitive skills, in interaction with dispositions and socio-emotional skills.
- Professional development includes exposure to visible routines, protocols and instructional practices, where teachers see in practice new forms of instruction or assessment.
- These programs rely on a mix of opportunities for learning situated in the context of the schools where teachers work.
- To support the intensive and sustained activities of professional development that these various programs advance, the organizations in charge build a range of partnerships with institutions outside of schools that contribute various types of resources.
- These programs see teacher practice as situated in specific organizations and social contexts, and in general adopt a whole-school approach, rather than helping individual teachers increase their capacity.
- These programs all develop capacities among teachers to advance pedagogies with the goal of developing competencies that are not formally assessed in the school or school system. The organizations that support these various programs all model a learning orientation.

1.7 A Role for Universities Developing and Implementing Contextually Appropriate Strategies for Climate Change Education

The specific impacts of climate change in particular communities and geographies differ, as do the ways in which particular communities contribute to climate change. As a result, the way in which people need to adapt to climate change, mitigate the impacts of climate change on their lives or diminish their impact on climate change need to be fit to context. For this reason, it is necessary that climate change education is situated in particular places, in particular geographies and economies. Whether the goal is to educate students for individual or collective actions, those have to be relevant to specific contexts. Generic lists of suggestions or teaching guides are of limited value to address these contextual specificities.

In addition, the institutional characteristics of education systems, schools and teachers differ in terms of their strengths and shortcomings, in some settings teachers are better educated than in others, some schools are institutionally weaker than others. Schools and systems also have different strengths, and any change process needs to build on those particular strengths. This wide variation in the context of schools limits the value of 'one size fits all' approaches to curriculum or teacher preparation for climate change education. While there is knowledge that is of universal value, such as knowledge about the science of climate change, translating that knowledge into a process that can develop particular competencies that matter to mitigate or adapt to climate change requires localization as does developing specific curriculum for particular teachers or particular students.

The transformation of institutional culture in order to effectively educate for climate change is too ambitious a task to be undertaken by schools or non-formal education organizations in isolation. They are too small to have the resources necessary to develop high quality curriculum and to, on their own, enhance the capacity of their teachers or facilitators to do this well. Such an effort requires a level of scientific expertise and innovation that calls for a larger scale and quality of resources than are typically available inside single schools. For example, there are many online resources that can be usefully deployed to teach about climate change, but curating these resources is one task that can take considerable time, for this reason assistance from university students in making thoughtful selections among the vast resources available in universities, and in the world wide web, to support teachers identifying suitable curriculum resources would be helpful.

It may also be unreasonable to ask of teachers that they single handedly take on the local or national politics which muddle the conversations about climate change with ideas that are extraneous to the scientific consensus. Engaging teachers in networks with other educators, and with colleagues in universities including scientists, can leverage the support of a professional community in developing teacher capacity to engage professionally with the topic.

If more scale is necessary to innovating in Climate Change Education than the scale available to single schools, why not leave the job to Ministries and Departments of Education? Because most departments of education have been designed to administer large systems, to manage resources efficiently and to ensure accountability, not to innovate. This is the reason most innovations in curriculum involve participation of other institutions and actors: disciplinary specialists, universities, professional associations.

Given the challenge of building teacher capacity, and given that Ministries of Education have proven that they are not well equipped to promote the necessary innovation to sustain high quality climate education at scale, where can the institutional resources to address this challenge be sought? A way to address this shortcoming in capacity is to repurpose an existing institutional resource in service of climate change education. That institutional resource are universities. Universities are ideally suited to take on this role of strengthening the capacity of schools and school systems to advance climate change education, translating the science of climate change, and the knowledge about what works for deeper learning, into specific programs that are tailored to concrete students, teachers and schools. Universities can do this in partnership with schools or non-formal education institutions that are within their immediate vicinity, or with schools that are situated more remotely if universities have ties of sufficient depth to those localities.

Clearly educating the public and supporting more sustainable choices requires also engagement of media, employers and governments. Universities, however, are unique in that they educate those professionals who work in those industries and others, potentially having a significant multiplier effect over time. Universities are also uniquely suited to educate about climate change because they contain, within the various departments of physics, chemistry, biology, the disciplinary expertise to develop high quality curriculum that reflects the current scientific consensus. They also have the scientific expertise to think about development comprehensively, and therefore to help students understand climate change in the broader context of poverty reduction and sustainability. Just as important, they also house, within their education departments, the expertise to translate the scientific knowledge base into effective k-12 curriculum and programs of teacher professional development. Drawing on collaboration across various disciplines, they can develop instructional resources that support the sophisticated thinking and action necessary for adaptation, mitigation and reversal of climate change. For instance, departments of public policy and social sciences can develop simulations that engage students in negotiating different views with respect to various policy proposals such as caps on emissions or carbon tax. As some of the competencies which students need to develop to think through options in response to climate change are ethical, there is also an important role for the humanities in cultivating the moral imagination of learners as they develop responses to climate change. Universities are the most capacious institutions for this kind of interdisciplinary collaboration in developing a rich curriculum and transformative pedagogies that support deeper learning about climate change.

Furthermore, the ubiquity of institutions of higher education around the world, makes it possible to advance efforts of curriculum development and capacity building which are responsive to the particular needs and opportunities of local contexts. An additional reason why universities are a strategic resource to support schools in addressing climate change is because they can integrate teachers into larger networks which provide them the intellectual resources and the support to understand that they are not alone in teaching about this topic. This support may be especially valuable in contexts where the politization of the subject of climate change challenges teachers.

An additional reason for universities to engage in this work, is that engaging in designing approaches to climate change education will teach university students how to solve problems and design innovations around climate change. Engaging students in problem-based learning, in specific contexts, helping them unravel the systems that undergird current climate predicaments in particular locales will help them learn to change them. In engaging them with local actors, this form of climate change education will also help students develop the skills for collaboration that will serve them well to continue to advance the necessary systemic change to curb climate change. These are the opportunities to gain the skills to understand and transform systems which a recent review of research on climate change and energy education is largely absent from most of the efforts they studied (Jorgenson et al. 2019).

Universities are already concerned with climate change and advancing a range of actions to address it, although more intentional and effective instruction is necessary in order for all university students to gain the necessary skills to understand, adapt to and mitigate climate change (Leal Filho and Hemstock 2019). A recent survey administered to university leaders and faculty working on sustainability in 51 countries reveals a gap between the priority universities attach to climate change and the generalization of opportunities for all students to learn about climate change (Leal Filho et al. 2019). Whereas 59% of the respondents agreed or strongly agreed with the statement that their university attaches strong priority to matters related to climate change, only 41% of them agreed or strongly agreed with the statement that there are opportunities to learn about climate change in the courses chosen by students, and 58% indicated that there are courses related to education for sustainable development available in the university curriculum. These responses indicate that climate change education is an option, but not a requirement, in many institutions of higher education (Leal Filho and Hemstock 2019, p. 7). Unsurprisingly, given that the respondents included faculty and senior staff working on sustainability, these responses about existing opportunities to learn about climate change contrast with the much higher number, 96% of respondents to the same survey, who agree or strongly agree with the statement that 'a university should encourage its students to search for solutions with regards to problems caused by climate change' (Leal Filho and Hemstock 2019, p. 7).

There are examples of the power of such collaborations between universities and schools to support more effective curriculum and instruction. In the United States, the new science standards were designed as a result of a collaboration between the National Research Council, the National Science Foundation, the American Association for the Advancement of Science and the National Science Teacher Association. In Chile, one of the most successful programs of inquiry-based science education in public schools was designed by science professors at the Universidad de Chile, inspired by a similar program developed by the Smithsonian Institution in the United States. In France, one of the leading organizations to support the improvement in the quality of science and technology teaching at the elementary levels, the Foundation 'La main a la pate' was established by the Academy of Sciences, and the Ecole Normale Superieure of Paris and of Lyon. The school of earth, energy and environmental sciences at Stanford University has designed a very high quality climate change education curriculum, which reflects the level of scientific rigor that results when scientists studying the topic translate what they know for lay audiences (Stanford University 2020). The National Aeronautics and Space Administration Agency has synthesized most of the scientific consensus, generated in universities, with respect to climate change (NASA 2020). Simply put, universities have unparalleled intellectual resources to develop climate change education curriculum and programs of teacher professional development that can support the necessary development of capacity that the enterprise demands. Furthermore, they have the know-how and the practice to experiment, evaluate, conceptualize and theorize which are essential to helping develop the field of climate change education into one where what is currently an largely undertheorized practice develops into a professional field of practice, into a field guided by expert knowledge and supported by the powerful tools of logic and science.

1.8 Development of the Approaches to Climate Change Education in This Book

The purpose of this book is to inspire university faculty, students and leaders to integrate climate change education into existing courses, in ways which provide students opportunities to design approaches to climate change education which could be implemented in schools and non-formal education institutions at the precollegiate level. Doing this would serve two purposes. One to educate university students themselves so they learn how to design solutions to existing problems, rather than merely contemplate them. This is of intrinsic value to students in higher education, regardless of the specific problems they learn to solve. It is an approach to support deeper learning and the development of twenty first century skills, and to help students develop hope and self-efficacy in tackling challenging problems. The fact that climate change is one of the most critical challenges of our times, makes it especially important to cultivate students' capacity to be change agents, rather than bystanders.

The other purpose this approach would advance is to augment the capacity of primary and secondary schools to educate about climate change, by relying on their partnership with universities. If only ten percent of the more than twenty thousand institutions of higher education around the world engaged only one professor each year in such an effort, each engaging five teams of students in the design of approaches to climate change education, this would produce each year 10,000 context specific strategies which could be offered to local coalitions for validation, adaptation, implementation and evaluation. The equivalent of 2,000 books like this one a year. This likely exceeds all the resources created by all international development institutions to support climate change education since the United Nations Sustainable Development Goals were adopted in 2015.

Over time, sustaining such efforts would build considerable expert knowledge about how best to educate communities to adapt to, mitigate and redress climate change. It is hard to imagine a more capacious engine than universities to design, research and support the development of capacity that this field needs with urgency in order to close the institutional capacity gap at the root of the climate education paradox. A production of 2,000 resource books a year systematizing climate change education would certainly contribute to theorizing the practice of climate change education and building this field of research and practice.

The next five chapters in this book exemplify context-specific education programs on climate change. They were developed in the context of a graduate course on education policy analysis I teach at the Harvard Graduate School of Education. The course covers the subject of policy analysis and design, as well as substantive themes focused on deeper learning, system level reform, curriculum, and teacher preparation. The course offers students the opportunities to work on several real-life projects, including consulting for governments and developing an approach to address climate change for a specific setting. To approach this task, students had to identify a real institutional partner for their work, an institution with an interest on the topic of climate change in a particular geographic setting. The students then studied the specifics of the particular challenge of climate change in that setting, the unique strengths and needs of the institution, and considered various alternatives to addressing the challenge (Bardach and Patashnik 2016). In effect, they addressed the key questions which I argue are essential to develop a context specific strategy of climate change education:

1. What are the specific impacts of climate change in this jurisdiction? How do they impact various human populations? How do human activities contribute to climate change?
2. What knowledge, dispositions and behaviors could mitigate the impact of climate change and are there ways in which changes in the behaviors of populations in this jurisdiction could slow down climate change? What kind of collective action could influence systems that contribute to climate change?
3. What are the means of delivery to reach each of the specific populations in this jurisdiction who needs to be educated on climate change?
4. What curriculum can help educate each population?
5. What role can the institution we are collaborating with play in advancing climate change education in that jurisdiction?

Students remained in communication with this institutional partner as they examined the root causes of the problem and designed and evaluated several approaches to addressing it, in an iterative process that involved loops of analysis-design-feedback over an entire semester. They then wrote a paper which described and analyzed the curriculum they had produced, conceptualizing their practice and integrating this work with the literature on climate change they had studied to support their design. Once their paper was completed, all authors of the chapters included in this book reviewed the entirety of the approaches developed to ensure greater coherence and alignment across the entire book, using this introductory chapter to guide those revisions. The papers were then presented at a global education conference at which leaders in the field of international development, including education specialists from UNICEF, the President of an international development organization and a former secretary of education of Mexico provided feedback to drafts of these chapters. They also received feedback to the programs they had developed from their partners and revised their chapters based on that first round of feedback. Once the full manuscript was finished, we received additional feedback from the editors of this series at Springer, Annette and Noel Gough, and from two anonymous reviewers, and made additional revisions to the chapters based on it.

The authors of these papers are educators with professional experience as teachers, child advocates, trainers, leadership mentors, curriculum developers and writers in Australia, Cambodia, China, Guatemala, Honduras, Japan, Nicaragua, Colombia, Pakistan, South Korea, Thailand, Uganda, United Kingdom, and the United States.

The programs they designed illustrate context specific climate change education programs, focusing on schools, non-formal settings and educator preparation institutions. The chapter are written with the aim of offering examples of general value beyond the specific contexts for which they were designed, but the essence of each chapter remains that, to be useful, climate change education needs to be firmly grounded in the specifics of a context and to be responsive to that context. We hope this level of detail, and the conceptualizations offered to justify the curricular and design choices which were made, will make these materials more useful to those who seek to adapt them to their particular contexts, or develop their own programs drawing inspiration from these examples.

Chapters 2 and 3 focus on climate change education in schools, each representing an approach to climate change education: a specific curriculum and a whole school approach to change. In contrast, Chaps. 4 and 5 focus on non-formal settings, Chap. 4 develops a specific skills-building radio education program whereas Chap. 5 develops a climate-change education skills program integrated into a larger literacy and life skills program in Pakistan. Chap. 6 focuses on preparation of graduate students in education, preparing for a variety of roles in 'whole of education system' reform.

In Chap. 2 'Learn to Lead: Developing Curricula that Foster Climate Change Leaders' David Rhodes and Margaret Wang focus on the question of how to develop leadership capacities that help students in Israel and Palestine address climate change. Their approach to curriculum development engages an institution of civil society that brings together science students from Israel and Palestine to work on climate related topics, in the process of supporting schools. The curriculum emphasizes deeper learning and the development of transferable skills, and not just the transmission of factual knowledge.

In Chap. 3 'Creating a Culture of Shared Responsibility for Climate Action in Guatemala through Education' Lina Lopez Lalinde and Carrie Maierhofer develop a whole school approach to climate change education in Guatemala. They explain why following UNESCO's school-based approach of whole-school transformation is sensible, in a context in which climate change education is already included in the curriculum but not really implemented because of inadequate capacity at the school level, and examine the requirements of such an approach to building teacher capacity, proposing ways to address them relying on partnerships between government and organizations of civil society.

In Chap. 4 'Rezistans Kimatik. Building Climate change resilience in Haiti through educational radio programming' Ashley Bazin and Christelle Saintis examine the challenge of responding to specific climate related challenges to vulnerable populations in Haiti, and opt for a non-formal radio-based approach to educate youth and adults.

In Chap. 5 'Adaptation, Migration, Advocacy. A Climate Change Curriculum for Out-of-School Children in Badin, Sindh' Natasha Japanwala develops a non-formal education program to educate youth in a region of Sindh, Pakistan, where traditional agricultural livelihood is challenged by salinization of land for adaptation to those changes and mitigation of their impact in their lives.

Finally, in Chap. 6 'Students as Partners. Implementation of a Climate Change Education within the Harvard Graduate School of Education' Annie Nam and Sueyoon Lee discuss the potential of student led curriculum development on climate change in schools of education, and develop a prototype of such a program for students at the Harvard Graduate School of Education.

In the concluding chapter I extract lessons and implications from these projects for future university based instruction of the sort I adopted in this course.

Our hope is that these five curricula will add to the ongoing knowledge base about climate change education, offering specific examples of how to advance this important area of education in ways which are responsive to context. We also hope that the entire book will illustrate the potential of engaging universities in designing climate change education curriculum, as a way to augment the capacity of schools, to develop necessary context specific approaches, and to educate university students to invent solutions to this defining issue of our times.

References

Alberta Council for Environmental Education. (2017). *What is excellent climate change education?* Alberta: Alberta Council for Environmental Education.

Arora-Jonsson, S. (2011). Virtue and vulnerability: Discourses on women, gender and climate change. *Global Environmental Change, 21*, 744–751.

Bardach, E., & Patashnik, E. (2016). *A practical guide for policy analysis. The Eightfold Path* (5th ed.). Thousand Oaks: CQ Press.

Berger, M. (2019). Italy's government becomes first to mandate climate change education in schools. *Washington Post*. November 6, 2019. https://www.washingtonpost.com/world/2019/11/06/thats-why-we-exist-italys-government-becomes-first-mandate-climate-change-education-schools/

Brueck, H. (2019). A $350 toilet powered by worms may be the ingenious future of sanitation that Bill Gates has been dreaming about. *Business Insider*. https://www.businessinsider.com/bill-gates-foundation-helps-invent-tiger-toilets-powered-by-worms-2019-1

Chen, G. (2017). Climate change to become part of core curriculum in public schools. Public School Review. https://www.publicschoolreview.com/blog/climate-change-to-become-part-of-core-curriculum-in-public-schools

D'Agostino, R. (2018). How does Bill Gates's ingenious, waterless, life-saving toilet work? https://www.popularmechanics.com/science/health/a24747871/bill-gates-life-saving-toilet/

Delors, J. (1996). *Learning: the treasure within, report to UNESCO of the International Commission on Education for the Twenty-first Century*. Paris: UNESCO.

Gibb, N. (2016). *Getting climate-ready: A guide for schools on climate action*. Paris: UNESCO. https://unesdoc.unesco.org/ark:/48223/pf0000246740.

Globe Scan. (2017). The 2017 climate survey. Evaluating progress on climate change. https://globescan.com/wp-content/uploads/2017/11/GlobeScan-SustainAbility-Survey-Evaluating-Progress-on-Climate-Change-Nov2017.pdf Accessed 1 Mar 2020.

Gough, A. (2013). The emergence of environmental education: a 'history' of the field. In R. B. Stevenson, M. Brody, J. Dillon, & A. Wals (Eds.), *International handbook of research on environmental education* (pp. 13–22). New York: Routledge for the American Educational Research Association.

Gough, A. (2016). Environmental sustainability in schools. In T. Barkatsas & A. Bertram (Eds.), *Global learning in the 21st century* (pp. 83–101). Rotterdam: Sense Publishers.

Gough, A. (2020). Educating Australia on the climate crisis. *Asia and the Pacific Policy Society Policy Forum*. 5 Feb 2020. https://www.policyforum.net/educating-australia-on-the-climate-crisis/

Gough, A., Lee J.C.K, and Tsang, E. (2020, in press) Green school movements: An introduction. In Gough, A., Lee, J.C.K, and Tsang, E.P.K. (Eds.) *Green schools globally. Stories of impact on education for sustainable development*. Cham: Springer.

Haas, R. (2020). *The world. A brief introduction*. New York: Penguin Press.

Inglehard, R., Haerpfer, C., Moreno, A., Welzel, C., Kizilova, K., Diez-Medrano, J., Lagos, M., Norris, P., Ponarin, E., Puranen, B., et al. (Eds.). (2014). *World values survey: Round six—*

Country pooled datafile 2010–2014. Madrid: JD Systems Institute. Version: http://www.worldvaluessurvey.org/WVSDocumentationWV Accessed 1 Mar 2020.

International Panel on Climate Change (IPCC). (2014) Climate Change 2014: Synthesis Report. *Contribution of Working Groups I, II and III to the Fifth Assessment Report of the Intergovernmental Panel on Climate Change* [Core Writing Team, R.K. Pachauri and L.A. Meyer (eds.)]. IPCC, Geneva, Switzerland, 151 pp.

IPCC. (2018). Summary for Policymakers. In: Global Warming of 1.5°C. An IPCC Special Report on the impacts of global warming of 1.5°C above pre-industrial levels and related global greenhouse gas emission pathways, in the context of strengthening the global response to the threat of climate change, sustainable development, and efforts to eradicate poverty [Masson-Delmotte, V., P. Zhai, H.-O. Pörtner, D. Roberts, J. Skea, P.R. Shukla, A. Pirani, W. Moufouma-Okia, C. Péan, R. Pidcock, S. Connors, J.B.R. Matthews, Y. Chen, X. Zhou, M.I. Gomis, E. Lonnoy, T. Maycock, M. Tignor, and T. Waterfield (eds.)]. World Meteorological Organization, Geneva, Switzerland, 32 pp.

Jain, P. (2020). Reforming educational systems with sustainability at their core. In UNESCO (2020), *Humanistic Futures of Learning. Perspectives from UNESCO Chairs and UNITWIN Networks* (pp. 27–30). Paris: UNESCO. https://unesdoc.unesco.org/ark:/48223/pf0000372577

Jorgenson, S. N., Stephens, J. C., & White, B. (2019). Environmental education in transition: A critical review of recent research on climate change and energy education. *The Journal of Environmental Education, 50*(3), 160–171.

Kamenetz, A. (2019). Most teachers don't teach climate change; 4 in 5 parents wish they did. National Public Radio. April 22, 2019. https://www.npr.org/2019/04/22/714262267/most-teachers-dont-teach-climate-change-4-in-5-parents-wish-they-did Accessed 1 Mar 2020.

Karrow, D., DiGiuseppe, M., & Inwood, H. (2020). Environmental and sustainability education in teacher education: Canadian perspectives. In D. Karrow & M. DiGiuseppe (Eds.), *Environmental and sustainability education in teacher education: Canadian perspectives* (pp. 2–18). Cham: Springer.

Leal Filho, W., & Hemstock, S. (Eds.). (2019). *Climate change and the role of education*. Cham: Springer.

Leal Filho, W., Mifsud, M., Molthan-Hill, P., Nagy, G., Veiga Avila, L., & Lange Salvia, A. (2019). Climate change skepticism at universities: a global study. *Sustainability, 2019*(11), 2981.

Mantatov, V., Mantatova, L. & Nasibulina, A. (2020). Humanism and environmental ethical frameworks. In UNESCO (2020). *Humanistic futures of learning. Perspectives from UNESCO Chairs and UNITWIN Networks* (pp. 34–38). Paris: UNESCO. https://unesdoc.unesco.org/ark:/48223/pf0000372577

Mayherfeld, M., & Askhood, L. (2015). *An invitation to environmental sociology*. Thousand Oaks: Sage Publications.

Ministerial Council on Education, Employment, Training and Youth Affairs. (2008). *Melbourne declaration on educational goals for young Australians*. http://www.curriculum.edu.au/verve/_resources/National_Declaration_on_the_Educational_Goals_for_Young_Australians.pdf

NASA. (2020). Global climate change. Vital Signs of the Planet. https://climate.nasa.gov/ Accessed 14 Jan 2020.

NOAA National Centers for Environmental Information. (2020). S*tate of the climate: Global climate report for annual* 2019. https://www.ncdc.noaa.gov/sotc/global/201913

OECD. (2012). How green are today's 15 year olds. https://www.oecd.org/pisa/pisaproducts/pisainfocus/50150271.pdf

OECD. (2019). PISA 2018 results (volume I): What students know and can do. *PISA, OECD Publishing, Paris*. https://doi.org/10.1787/5f07c754-en.

OECD. (2020). Conceptual learning framework. Core Competencies for 2030. http://www.oecd.org/education/2030-project/teaching-and-learning/learning/core-foundations/Core_Foundations_for_2030_concept_note.pdf. Accessed 5 May 2020.

Ojala, M. (2012). Hope and climate change: the importance of hope for environmental engagement among young people. *Environmental Education Research, 18*(5), October 2012, 625–642.

Ojala, M. (2016). Facing anxiety in climate change education: From therapeutic practice to hopeful transgressive learning. *Canadian Journal of Environmental Education, 21*, 41–56.

Pidcock, R. (2015). Polar bears and climate change. What does the science say. Carbon brief clear on climate. https://www.carbonbrief.org/polar-bears-and-climate-change-what-does-the-science-say

Plutzer, E., Lee Hannah, A., Rosenau, J., McCaffrey, M. S., Berbeco, M., & Reid, A. H. (2016). *Mixed messages: How climate is taught in America's schools*. Oakland: National Center for Science Education. http://ncse.com/files/MixedMessages.pdf

Poullaouec-Gonidec, P. (2020). Reinventing the world through landscape reading. In UNESCO (2020). *Humanistic Futures of Learning. Perspectives from UNESCO Chairs and UNITWIN Networks*, (pp. 31–33). Paris: UNESCO. https://unesdoc.unesco.org/ark:/48223/pf0000372577

Reimers, F. (2007). Alberto Vinicio Baez and the promotion of science education in the developing world. *Prospects, 37*(3), 369–381.

Reimers, F. (2020a). *Audacious education purposes. How governments transform the goals of education systems*. Cham: Springer Nature.

Reimers, F. (2020b). *Educating students to improve the world*. Singapore: Springer.

Reimers, F. (2020c). *Implementing deeper learning and 21st education reforms. Building an education renaissance after a global pandemic*. Cham: Springer Nature. In Press.

Reimers, F., Adams, R., & Berka, M. (2019). *Creative lessons to open classrooms and minds to the world*. Washington, DC: The NEA Foundation.

Reimers, F., Barzano, G., Fisichella, L., & Lissoni, M. (2018). *Cittadinanza globale e sviluppo sostenibile*. Milan: Pearson.

Reimers, F., & Chung, C. (Eds.). (2016). Teaching and learning for the twenty first century. Cambridge: Harvard Education Publishing.

Reimers, F., & Chung, C. (Eds.). (2018). *Preparing teachers to educate whole students: An international comparative study*. Cambridge: Harvard Education Publishing.

Reimers, F., et al. (2017). *Empowering students to improve the world in sixty lessons*. Charleston: CreateSpace.

Reimers, F., Adams Jr, R., Shannon, K. (2018). *Twelve lessons to open classrooms and minds to the world*. Charleston: Kindle Direct Publishing.

Rousell, D., & Cutter-Mackenzie-Knowles, A. (2020). A systematic review of climate change education: giving children and young people a 'voice' and a 'hand' in redressing climate change. *Children's Geographies, 18*(2), 191–208.

Saad, L. (2019). Preference for environment over jobs largest since 2000. https://news.gallup.com/poll/248243/preference-environment-economy-largest-2000.aspx

Stanford University. (2020). Climate change education https://earth.stanford.edu/climate-change-ed

Stevenson, R. B., Nicholls, J., & Whitehouse, H. (2017). What is climate change education? *Curriculum Perspectives, 37*, 67–71.

Stewart, I. (2020). Geology for society: Earth science for sustainable development. In UNESCO (2020). *Humanistic Futures of Learning. Perspectives from UNESCO Chairs and UNITWIN Networks*, (pp. 39–41). Paris: UNESCO. https://unesdoc.unesco.org/ark:/48223/pf0000372577

UN Women. (2016). Leveraging co-benefits between gender equality and climate action for sustainable development. Mainstreaming gender considerations in climate change projects. https://unfccc.int/files/gender_and_climate_change/application/pdf/leveraging_cobenefits.pdf

UNCED. (1992). United Nations Conference on Environment and Development (Earth Summit) (3–14 June 1992, Rio de Janeiro, Brazil) https://www.un.org/en/events/pastevents/UNCED_1992.shtml

UNDP. (2013). Gender and climate change – Asia and the Pacific. Policy Brief 1. Overview of linkages between gender and climate change. 10 Dec 2013. https://www.undp.org/content/undp/en/home/librarypage/womens-empowerment/gender_and_environmentenergy/gender-and-climate-change-asia-pacific.html

UNEP. (2012). *Global environment outlook – 5: Environment for the future we want*. Nairobi: United Nations Environment Programme.

UNESCO. (2010). Climate change education for sustainable development: The UNESCO climate change initiative. https://unesdoc.unesco.org/ark:/48223/pf0000190101

UNESCO. (2016). *Education for people and planet: Creating sustainable futures for all*. Paris: UNESCO. https://unesdoc.unesco.org/ark:/48223/pf0000245752

UNESCO. (2018). Progress on education for sustainable development and global citizenship education: findings of the 6th consultation on the implementation of the 1974 recommendation concerning education for international understanding https://unesdoc.unesco.org/ark:/48223/pf0000266176

UNESCO. (2019a). Country progress on climate change education, training and public awareness. An analysis of country submissions under the United Nations Framework Convention on Climate Change. https://unesdoc.unesco.org/ark:/48223/pf0000372164

UNESCO. (2019b). *Educational content up close*. Paris: UNESCO. https://unesdoc.unesco.org/ark:/48223/pf0000372327.

UNICEF. (2017). Reimagining life skills and citizenship education in the Middle East and North Africa: A four-dimensional and systems approach to 21st century skills conceptual and programmatic framework. https://www.unicef.org/mena/media/6151/file/LSCE%20Conceptual%20and%20Programmatic%20Framework_EN.pdf%20.pdf

United Nations. (1992). United Nations Framework Convention on Climate Change. https://unfccc.int/resource/docs/convkp/conveng.pdf

United Nations. Framework convention on climate change. (2014). Conference of the Parties. The Lima Ministerial Declaration on Education and Awareness-raising. https://unfccc.int/resource/docs/2014/cop20/eng/l01.pdf

United Nations. (2020). Sustainable development goals knowledge Platform. https://sustainabledevelopment.un.org/sdg4

Whitehouse, H. (2019). Ever wondered what our curriculum teaches kids about climate change? Not much. *The Conversation*. 28 Sept 2019. https://theconversation.com/ever-wondered-what-our-curriculum-teaches-kids-about-climate-change-the-answer-is-not-much-123272

World Economic Forum. (2020). The global risks report 2020. http://www3.weforum.org/docs/WEF_Global_Risk_Report_2020.pdf Accessed 1 Mar 2020.

Worldometers 2019. https://www.worldometers.info/world-population/world-population-by-year/

Fernando M. Reimers is the Ford Foundation Professor of the Practice of International Education and Director of the Global Education Innovation Initiative and of the International Education Policy Masters Program at Harvard University. An expert in the field of Global Education, his research and teaching focus on understanding how to educate children and youth so they can thrive in the twenty-first century. He is a member of UNESCO's Commission on the Futures of Education.

He has written or edited over a hundred academic articles and chapters and thirty-four books, of which the most recent, focusing on system level change to make education more relevant, include: Implementing Deeper Learning and 21st Education Reforms. Building an Education Renaissance after the COVID-19 Pandemic; Educating Students to Improve the World; Audacious Education Purposes. How governments transform the goals of education systems; Empowering teachers to build a better world. How six nations support teachers for twenty-first century education; Letters to a New Minister of Education; Teaching and Learning for the 21st Century; Preparing Teachers to Educate Whole Students: An International Comparative Study.

Chapter 2
Learn to Lead: Developing Curricula that Foster Climate Change Leaders

David Rhodes and Margaret Wang

2.1 Introduction

Imagine a group of ninth-grade students on a class trip, walking around the neighborhood of their school. The school is called Al Taqwa, and it is located in Rahat, a predominantly Bedouin city in the Negev desert of southern Israel. As the students explore the neighborhood, their assignment is to pay attention to any environmental problems they notice and take pictures of evidence related to those problems. Some students smell the toxic scent of burning plastic and they notice smoke rising from a pile of trash near the street. They snap several photographs before continuing on.

After returning from the trip, the teacher facilitates a discussion that engages the students in analysing the causes of the problems they observed. In the example of the burning garbage, the students explore how they might personally contribute to the problem by purchasing food and other items with plastic packaging. Then, they expand their analysis to consider the functions of the local and national government in managing waste. Questions are also raised with regard to the role of culture, highlighting the fact that burning waste was a common practice in Bedouin society with a history that extends back to a time when all the waste was organic. As the web of cause and effect grows increasingly complex, the class begins to incorporate causal diagrams to organize the information, and they start to construct maps of the different stakeholders involved.

In this course, the goal is to enable students to conceptualize and help address climate change problems in the region – problems associated with heat waves, droughts, dust storms, soil degradation, changing precipitation patterns and dryland water scarcity. The teachers, Fareed Mahameed and Samah Matany, are incorporating our high school Climate Change Leadership curriculum to equip and motivate students to affect change.

D. Rhodes · M. Wang (✉)
Harvard Graduate School of Education, Cambridge, MA, USA

F. M. Reimers (ed.), *Education and Climate Change*, International Explorations in Outdoor and Environmental Education,
https://doi.org/10.1007/978-3-030-57927-2_2

The Climate Change Leadership curriculum is designed to integrate disciplinary concepts and process-based skills associated with defining problems, gathering evidence, generating solutions, synthesizing the possibilities for action, drawing conclusions with regard to the best way forward, communicating the results of the process and taking action. Throughout the curriculum, there are opportunities for students to practice empathy, systems-thinking, media literacy, and collaboration as they engage in projects connected to climate change. Clear standards, units, and lesson plans combined with an accompanying project-based assessment open possibilities for teachers to integrate the curriculum in a wide variety of ways.

We recognize that, in order for a curriculum to be effective, it needs to be accompanied by a strong implementation plan, and we are currently working with educators like Fareed to develop these plans for a variety of contexts. Ultimately, this approach to innovation in education is operating from a professional perspective that focuses on supporting the development of the capacity of teachers and school leaders to affect change (Reimers 2020). If we can build a network of teachers and school leaders interested in implementing the Climate Change Leadership curriculum and we can identify the customized support necessary for successful implementation, then the participating teachers will be able to enable climate change leaders in the new generation. We specifically focus on the development of a particular curriculum and its implementation because, in the context of Israel, there have been 17 reform initiatives in the past 30 years. If the curriculum were connected to a large scale reform, it could be more likely for it to be implemented in schools only so long as the initiative lasts, undermining the potential for longer-term integration of the curriculum.

In the first part of this chapter, we identify the problems in secondary school climate change curricula, particularly in Jordan, Israel, and Palestine. Then, we analyze the role of tertiary institutions through their relevant climate change pedagogical practices. Finally, we summarize the effective secondary education climate change resources that inform our curriculum. In the second part of the chapter, we present a Climate Change Leadership curriculum, highlighting how it integrates problem-solving frameworks from across disciplines to inspire inquiry and thoughtful action. We then describe our proposed standards to justify the pedagogical principles that are the basis of our curriculum. Lastly, we elaborate on our context-specific approach to the development of implementation plans to help schools successfully incorporate the curriculum.

We created this curriculum in collaboration with the Arava Institute, a tertiary education program in southern Israel that brings together Israeli, Palestinian, Jordanian and international students from outside the region to engage in multidisciplinary environmental studies, encompassing natural and social sciences as well as research into cross-border environmental issues. Thus, this curriculum was contextualized for Israel, Palestine, and Jordan, and we sought to learn from the work of the Arava Institute at the higher education level as we designed the curriculum for secondary schools in the region. While the regional focus has guided choices related to some of the curricular content, it is important to note that school leaders or teachers in diverse contexts could use the curriculum as well. Evidence that our project is not specific to the regional context can be seen in the success of Eco-Schools – a global Education for Sustainable Development program with an approach to climate change education that aligns with core aspects of our curriculum (Andreou 2020). Ultimately, our goal is to help shift the conversation of climate change curricula away from how we teach climate change to how we can foster environmental leaders within younger generations.

2.2 Analysis of Climate Change Curricula

2.2.1 The Problem with Climate Change Curricula in Israel, Jordan, and Palestine

In Jordan, Palestine, and Israel, the governments have acknowledged the need for enhanced climate change education due to the region's semi-arid climate. More specifically, climate projections for the region include a rise of 3 °C (4 °C in Jordan), 10–30% decrease in precipitation (60% for Jordan), seawater level rise of 10 mm leading to land loss, and an increase in extreme weather events including floods, droughts, and heat waves (EcoPeace Middle East 2019). These environmental fluctuations threaten food, energy, and national security. The impacts of climate change are further amplified by the region's influx of refugees and inter-state conflicts.

In this particular region, we chose to focus on a formal education initiative due to the sheer number of students in secondary schools. This may not be the case in many other regions as explored in Chap. 4, Haiti, and Chap. 5, Pakistan. As such, curriculum can be a powerful tool in implementing change in the formal education space and thus, affecting a large student population. In Israel, the secondary school net enrollment rate is around 98% with around 820,000 students in the formal secondary system; in Jordan, the rate is 80.86% with around 790,000 students; in Palestine, the rate is 50.86% with around 750,000 students (UNESCO Institute of Statistics 2020).

Furthermore, there is already some sort of precedence of teaching about climate change within the formal education system. In the Jordanian high school curriculum, "Earth Sciences and Environment" is a separate subject studied by most public high school students that combines knowledge, skills, and behavioral domains. The curriculum also teaches students how to collect environmental information in their community and integrate it into civic action campaigns for awareness and advocacy. However, while the textbooks give instructions regarding how to apply environmental literacy to informed action, teachers and students do not necessarily have the knowledge and skills to conduct these projects (Al-Newashi 2002, p. 48). The formal high school system in Palestine faces comparable issues. Palestine also has environmental education as a mandatory subject between seventh and tenth grade. Nevertheless, an in-depth analysis of their main environmental education topic "The Science of Health and the Environment" reveals that while the curriculum addresses awareness and investigative skills, it is lacking in building major ecological concepts and applying their knowledge through positive environmental actions (Karama 2016, p. 14–17).

In contrast to mandatory environmental education in Palestine, in the Israeli high school system, education related to the environment is not compulsory and is only offered to students who major in "Environmental Sciences" or "Earth and Environmental Sciences." The "Environmental Sciences" curriculum has not been significantly updated since 1983 and it is mostly focused on students studying ecology. While students do have the option of selecting environmental ethics or environmental planning and management as a study unit, this constitutes a small population (Sagy and Tal 2015, p. 69). On the other hand, the "Earth and Environmental Sciences" curriculum is more aligned with climate change education as it focuses on "developing environmental insight" through systems-thinking (Sagy and Tal 2015, p. 70). However, since it is a new course, few students are exposed to this style of environmental education. In total, less than 10% of Israeli high school pupils are exposed to environmental education (Sagy and Tal 2015, p. 71). Since 2004, there have been initiatives spearheaded by the Ministry of Education and the Ministry of Environmental Protection to certify "Green Schools" based on a review of environmental practices of the school and environmental education curriculum. However, these initiatives fall short because there is no ongoing evaluation after the certificate is awarded to ensure that the school continues to perform to the relevant standards (Tal 2020).

Outside of the formal curriculum, there have been initiatives by various schools to transform the way climate change is being taught in the region. This is derived from a larger global movement through the United Nations Educational, Scientific, and Cultural Organization (UNESCO). This organization has provided climate change educational resources that clearly show education's role in mitigating climate change: enabling the youth to take collective action. For example, in UNESCO's course for secondary school teachers on climate change education called *Climate Change in the Classroom*, there is an emphasis on "motivating action through local learning" and "motivating learners through real life experience" (Selby and Kagawa 2013, p. 8). UNESCO has worked with the One UN Climate Change Learning Partnership (UN CC: Learn 2013) to directly disseminate curricular resources embodying these ideals.

On a global scale, climate change education is moving towards project-based learning, which is a pedagogical practice where students engage with an authentic problem or question. Through inquiry and reflection, they ultimately create a public product as a summation of their investigation. Similar approaches can be seen in regional initiatives such as Youth Xchange, ASPnet, and South Eastern Mediterranean Environment Project (SEMEP). Youth Xchange is a collaborative initiative of UNESCO and the United Nations Environment Programme that has designed materials to enable educators to teach sustainable lifestyles throughout the Middle East, including schools in Jordan (UNESCO and United Nations Environment Programme 2011). Similarly, ASPnet is dedicated to enabling schools to engage in climate change education that includes multi-disciplinary, collaborative projects (UNESCO

Associated Schools Network n.d.). Resources include guidelines to build climate change education networks within schools and materials designed to facilitate climate change education across subject areas. Lastly, the South Eastern Mediterranean Environment Project (SEMEP) programs focused on teaching students to apply scientific methodology to address specific issues in their local contexts (Cyprus Ministry of Education and Culture 2012). For example, in Israel, Palestine, and Jordan, secondary school projects were developed related to water filtration and conservation. In Jordan and Palestine, students were involved in other SEMEP initiatives, including expanding green spaces to combat desertification, and exploring questions of ethics and global warming (Cyprus Ministry of Education and Culture 2012).

Through a review of the formal high school curricula in Israel, Jordan, and Palestine, there is clearly a need for updated curricula related to climate change combined with sufficient educator capacity to deliver the curricula. One way forward might be to adopt project-based initiatives in the region, inspired by UNESCO, to guide such curricula as these initiatives can help students learn about climate change and apply their knowledge in the context of local collaborative projects. In parallel with these initiatives, there would also be a need for substantive professional development for teachers to ensure that they have the skills and confidence to teach in this way. While this shift in climate change curricula and teacher training could be impactful, the impact would be limited by the fact that project-based initiatives are less attentive to ensuring that students come away with a deep understanding of frameworks for informed actions outside of the context of the curriculum and their particular projects. This undermines the ability of students to transfer their understanding and limits their ability to engage with the multi-faceted dimension of climate change outside of the context of the course. Therefore, in order to teach students how to translate their knowledge into *sustained* civic engagement, it is essential that the students not only engage in projects within their communities, but that they also deepen their understanding of the process-based underpinnings of informed civic action.

2.2.2 *Learning from Tertiary Level Climate Change Pedagogy in the Region*

The role of tertiary institutions extends beyond educating young adults – in the context of climate change education, they can be a source of knowledge for content and pedagogical practices. As such, we turned to the tertiary institutions in the region to analyze effective climate change curricula. For example, at the Technion in Israel, Daniel Orenstein teaches a course on environmental policy in which students develop their understanding of environmental policy analysis through the study and application of ideas from the work of Eugene Bardach's (2011) *A Practical Guide for Policy Analysis: The Eightfold Path to More Effective Problem*

Solving (The Eightfold Path) and Steven Cohen's (2014) *Understanding Environmental Policy*. The students also engage in environmental projects in collaboration with local NGOs. In an interview, Orenstein shared the underlying rationale for the structure of his course as follows:

> The [project] component is due to the desire to work closely with stakeholders to suggest real and implementable solutions to environmental challenges with local urgency, and to have a "built-in" group of individuals and organizations who are interested in the students' results. The [policy analysis] component is because I think Cohen's framework helps students build a very strong and interdisciplinary foundation and understanding of the given environmental challenge, but lacks operational steps to define policy alternatives and potential solutions, while the Bardach framework is void of depth of analysis of the given challenges, but provides a very well-developed framework for problem-solving. Together they compensate for each others' weaknesses (Orenstein, D. (2019, December). Email with authors).

In the Arava Institute, our partner tertiary institution, the Environmental Leadership seminar provides two tracks that "highlight the interplay between individual action and large scale institutional change as two important and complementary tactics for finding real solutions to today's environmental crisis" (Haber 2019, p. 1). The Sustainable Living Track involves exploration of a topic along with a hands-on project on campus with a facilitator and mentorship. The Environmental Diplomacy and Entrepreneurship Track includes activism and community organization through an international non-profit called WeSea as well as collaboration with business mentors. Similar to the UNESCO initiatives in secondary schools throughout the region, there is a focus on inquiry and experiential learning. However, in these courses, there is an emphasis on the assessment of the process alongside the public product in order to develop leadership skills and problem-solving strategies.

Thus, at the tertiary level, there are courses that enhance the civic agency of students by cultivating a deeper grasp of the disciplinary content as well as understanding and implementation of the process of policy analysis. But this approach to education related to climate change is not accessible at the secondary level, and only the students who can access higher education and then gravitate towards classes on environmental policy are exposed to this pedagogy. This clearly indicates a need for institutions of higher education and secondary schools to develop strategies that would enhance the exchange of ideas in ways that would make effective climate change education accessible to all.

2.2.3 *Effective High School Climate Change Education Resources*

Lastly, we reviewed high school climate change education resources outside of the region that would help inform climate change curricula.

2.2.3.1 Paleontological Research Institution: The Teacher-Friendly Guide to Climate Change

The scales involved in climate change research lie outside the range of human experience, spanning billions of years and involving the interactions of microscopic particles and waves (Zabel et al. 2017). The systems and scales associated with climate change are often embedded in different disciplines within and beyond the context of the sciences, including biology, chemistry, atmospheric science, oceanography, geology, environmental sciences, etc. (Ledley et al. 2017, p. 6). In the United States, the Next Generation Science Standards (NGSS) recognizes the inherent interdisciplinary nature of climate change by promoting "cross-cutting concepts." Yet, by focusing on scientific complexity, teachers and students would get inundated with content, which could perpetuate the notion that only specialists can truly understand the issues, and thus, only specialists can address climate change (Gonzalez-Gaudiano and Meira-Cartea 2012, p. 15). The combination of these factors renders it difficult to make the relevant content accessible in a high school curriculum.

As a result, the Paleontological Research Institute (PRI) created *The Teacher-Friendly Guide to Climate Change* (TFG) (Zabel et al. 2017) to address these obstacles by explicitly engaging with questions of cognitive bias and providing tools for systems thinking. With accessible and engaging language combined with models and images, the challenge of grappling with the systems and scales connected to climate change is demystified. Once the challenge is clear, the next step is to practice actually grappling with complex systems and models, and that can happen through the development of lessons and activities incorporated into a curriculum.

Overall, there are two ways in which the TFG mitigates the risk of conveying that only professionals can engage with the issue of climate change. First, it makes the essential foundation in science accessible without going into all of the details embedded in each discipline. Students can learn about the carbon cycle, for example, without learning the underlying biological processes associated with respiration. While the foundation in science is included in the TFG, there is also a recognition of the fact that this foundation itself can still feel overwhelming, especially in combination with the myriad risks associated with climate change. Thus, it is essential to engage with the psychological aspect of the challenge as well, and the TFG does just that, calling attention to the sort of big picture thinking that can help students assimilate information while maintaining a sense of motivation and hope.

2.2.3.2 Climate Interactive

Climate change involves multilayered systems with feedback loops and evolving dynamics. This presents a significant challenge to educators who seek to engage students in complex systems analysis and enable them to affect change (Ledley et al. 2017). In education, when real-life context is introduced through inquiry and experiential learning, systems thinking is often emphasized as a valuable process-based skill. Systems thinking can be defined as a practice for understanding complex and dynamic systems and can include modeling to help with decision making and implementation (Ledley et al. 2017, p. 6).

Climate Interactive is a think tank from Massachusetts Institute for Technology that directly tackles the complexities of systems thinking by providing video lesson resources and simulations to teach people system dynamics modeling in climate change issues. For example, through games like World Climate, students learn the complexity of climate change negotiations. World Climate is a simulation game, in which participants role-play as delegates to work together and create an international global agreement to combat climate change. Iterative drafts of the agreement are coupled with a real-time display of its effects, which are backed by robust scientific data. In an evaluation involving participant surveys and focus groups, 85% of participants increased their motivation to take action and begin the inquiry phase. Furthermore, the program taught major systems thinking insights (Climate Interactive 2019).

Climate Interactive has also released an online course called the Climate Leader which trains students in systems thinking to motivate global response to climate change. Topics such as areas of high leverage and feedback loops give students specific heuristic models to understand the effects of human activity on the environment and vice versa.

2.2.3.3 Project Look Sharp: Media Literacy

There is no agreed upon resource with the full body of knowledge related to the issue of climate change. While the vast majority of scientists (more than 97%) agree that human action impacts the climate and anthropogenic CO_2 is causing global warming, a skeptic could still point to the small number of scientists who deny that such a causal link exists (Zabel et al. 2017). Furthermore, the issue of climate change has social and political dimensions that add layers of complexity to the analysis of sources and credibility. Some individuals or groups may present information to serve an agenda that has nothing to do with the pursuit of knowledge or the wellbeing of society at large. The obstacle to understanding presented by the intentional manipulation of information is compounded by the fact that anyone can fall prey to misinformation or unconscious biases (or both). To mitigate these risks, it is essential for the consumer of information to have the skills to decode bias, analyze credibility, and identify important missing perspectives. This includes metacognitive reflection on one's own biases in order to approach questions with an open mind and seek to identify and fill blind spots. For educators, it can be a difficult process to facilitate the development of media literacy skills necessary to explore questions associated with climate change.

Project Look Sharp (n.d.), a media literacy institute at Ithaca College, has honed the process of constructivist media decoding (CMD) to address the need for students to build media literacy skills. The process of CMD is inquiry based, and the role of the teacher is to present rich media documents for the class to analyze. A rich media document is one that has layers of complexity that connect directly to media literacy goals while engaging with content relevant to the curriculum. The teacher can lead the decoding by asking questions that engage the students in deep analysis of content, bias, credibility, purpose, techniques, and impacts. Decoding also involves analysis of how the interpretations of the document can vary, illuminating the ways in which our personal biases can shape the way we interpret the world. It is important to note that the process of CMD should be incorporated into analysis of documents that teach content, but the purpose of the decoding of the media document is not to teach the content itself. In CMD, questions are not designed to lead students to the teacher's predetermined conclusions, but rather to build the skills of inquiry and analysis, empowering the students to engage with each other and the documents. When students share insights and raise important questions, they engage in the practice of drawing evidence-based conclusions. Ultimately, the students develop the skills of decoding and the habits of inquiry that they can apply in the context of any complex question in order to understand how to draw informed conclusions based on an analysis of evidence from diverse perspectives and sources.

2.3 Climate Change Leadership Curriculum

2.3.1 Rationale of the Curriculum

In our analysis of existing climate change curricula in Israel, Jordan, and Palestine, we identified a pressing need for a curriculum that not only provides the necessary foundation in the sciences, but also helps students understand how to interpret scientific information to draw their own informed conclusions; a curriculum that not only provides students with an experience of agency through a hands-on project, but that also ensures that they understand how to design a project with attention to complex power structures, interconnected systems, and the needs and interests of diverse stakeholders; a need for a curriculum that not only instills a sense of civic responsibility, but that also provides a deep understanding of how to fulfill that responsibility by setting clear goals and collaborating with others to affect change. In other words, there needs to be curricula that help the student internalize the process of creative problem-solving and informed action. Students should graduate from high school with the skills and inclinations to mitigate the risks of climate change. This cannot happen solely through the practice of science in the classroom, nor can it happen solely through service connected to a project in the school or community. While these experiences and the associated learning outcomes may be necessary, they do not add up to the capacity for thoughtful policy analysis, innovative design thinking, or practical understandings of how to engage across divides to transform the future by mitigating the risks of climate change.

As a result, we have created an example of what a climate change education could look like called The Climate Change Leadership Curriculum (see Appendix A). It is important to note that implementation of a project-based curriculum without educator and school leader capacity would be insufficient, as ascertained by the project-based Jordanian curriculum. The resources we are developing, including this curriculum, are therefore constantly in revision through feedback from educators and schools involved in implementation. Before elaborating on the implementation plan, we will explain the different components in our curriculum that are informed by the gaps identified in climate change education as well as inspired by tertiary institutions in the region and effective global secondary school resources mentioned earlier.

2.3.1.1 Anchor Skills

The anchor skills delineate what students need to be able to do in order to exercise leadership specifically in the context of climate change. We identified four overarching themes: empathy, systems thinking, media literacy, and collaboration. These anchor skills are embedded throughout the curriculum, and metacognitive reflection on their application helps to ensure that students gain the ability to transfer the skills to new contexts.

Based on the review of effective high school climate change curriculum, these anchor skills are standards that must be taught in alignment with the process-based skills that guide the student leaders, which will be discussed in the next section. For example, in order to define problems, students need empathy to be able to approach questions with humility and seek to understand diverse perspectives. In order to gather evidence, students need media literacy to understand how to assess the credibility of information and consider the ways in which bias can lead to blindspots that need to be filled by seeking out missing perspectives. In order to generate solutions, students need systems thinking to understand how complex systems can be impacted by affecting change in areas of high leverage. And in order to draw conclusions, students need to understand how changes to specific areas of high leverage can help to transform harmful dynamics, effectively and efficiently promoting positive change. Eventually, students need to understand how to collaborate and motivate groups to take action based on their conclusions and inspire others to do the same. Then, they need to maintain their curiosity as they iterate the process, taking into account the actual impacts of their actions.

2.3.1.2 Process-Based Skills

Whereas the anchor skills revolve around climate change leadership competencies, process-based skills outline what the students actually have to do to engage in effective problem-solving that leads to informed civic action and, additionally, what it means to do these steps effectively. The process-based skills are drawn from three essential resources: Bardach's (2011) *Eightfold Path*, Stanford Design School's design thinking process (Platner n.d.), and the National Council for the Social Studies' framework called College, Career, and Civic Life (C3) (2017).

If a goal of education is to foster thoughtful and engaged citizens, then policy analysis would be essential. However, the skills of the policy analyst are not explicitly included in traditional high school curricula, leaving the concept of policy analysis vague or inaccessible. In the *Eightfold Path*, public policy professor Eugene Bardach (2011, xvi) demystifies the conceptual framework behind policy analysis by identifying eight fundamental steps of the process, which we summarize below:

1. Define the problem – What is the problem under investigation?
2. Gather evidence – What perspectives and data would be necessary to analyze the way forward?
3. Construct alternatives – What are possible paths forward based on the evidence?
4. Select criteria – How should the alternative paths be weighed relative to one another? On what basis?
5. Project the outcomes – What would be the outcomes of the different paths?
6. Confront tradeoffs – What are the pros and cons of the projected outcomes relative to the criteria?
7. Focus narrow deepen decide – What path forward makes the most sense following an analysis of the tradeoffs?
8. Tell your story – What are effective ways to communicate the results of this process to garner support for the conclusion?

Similarly, design thinking brings in another layer of how to creatively think through problems in the following steps as summarized below (Platner n.d.):

1. Empathize – What does the target population care about and need?
2. Define – Based on the needs and interests of the target population, how should the problem be framed?
3. Ideate – What is a broad range of possible ways forward to address the problem?
4. Prototype – How can a prototype be built to further explore the best way forward?
5. Test – How can ideas for revision of prototypes be gained through testing and feedback?
6. Revise.

Lastly, the C3 framework also outlines process-based skills that begin with teaching students how to design inquiries and ends with communicating conclusions to take informed actions. Although this inquiry-based framework uses disciplinary concepts limited to social studies, the format could easily integrate science disciplines.

1. Dimension 1: Developing Questions and Planning Inquiries
2. Dimension 2: Applying Disciplinary Tools and Concepts (Civics, Economics, Geography, and History)
3. Dimension 3: Evaluating Sources and Using Evidence
4. Dimension 4: Communicating Conclusions and Taking Informed Action (National Council for Social Studies 2017)

The process based skills in our curriculum are structured following the sequence of Eugene Bardach's (2011) *Eightfold Path*, and they integrate the ideas embedded in design thinking and the C3 framework. In addition, the accompanying project directions for students outline these process-based skills in sequential order. The benefit of making this process transparent to students is that they can learn how to apply skills of policy analysis, design thinking, and inquiry to multiple aspects of climate change and/or other issues, and they could engage in a self-directed iterative process. In other words, by including the following process-based skills, students are able to internalize effective problem-solving processes:

1. Defining problems: developing compelling questions that promote inquiry and crafting meaningful and actionable problem statements.
2. Gathering evidence: Determining the types of relevant sources, evaluating for credibility, and gathering information from these various and diverse sources.
3. Generating solutions: Synthesizing evidence to generate a wide range of ideas and creating externalized prototypes to communicate potential solutions and receive feedback.
4. Concluding: Applying selection criteria, assessing strengths and weaknesses of outcomes, and drawing conclusions on how to proceed.
5. Communicating and acting: Communicating the results and trade-offs and taking direct actions towards implementing the results of the analysis.

This draws upon evidence of the efficacy of an inquiry learning cycle that fosters the students' sense of responsibility and autonomy in the process of knowledge generation and cultivates the ability to apply knowledge to new context (Cairns 2010). In many ways, our project builds upon work being done in environmental education grounded in experiential learning theory. For example, the Enviroschools Programme in New Zealand incorporates a learning cycle that involves identifying a current situation, exploring alternatives, taking action, and reflecting on change (Eames and Mardon 2020).

2.3.1.3 Disciplinary Tools and Concepts

All of the above skills are directly relevant to mitigating the risks of climate change, and they must be taught and assessed in order for students to emerge with the capacities for climate change leadership. But they will only lead to an impact on climate change if they are also utilized to help students understand what climate change actually is. We integrated disciplinary tools and concepts into our curriculum because leadership cannot be taught in the absence of hard knowledge. In other words, science is necessary in climate change leadership. Instead of replacing the science behind climate change that is usually taught in isolation, our curriculum provides an example of what it would look like for students to have the space to apply scientific knowledge. Thus, teachers must facilitate students' acquisition of the foundational disciplinary tools and concepts to navigate questions related to climate science.

TFG (Zabel et al. 2017) distills these tools and concepts in a clear conceptual framework, identifying the bigger ideas in earth system sciences necessary in understanding climate change. This includes the parts in the Earth System, energy processes, human environmental interactions, physical and chemical principles, and maps and models. In collaboration with PRI, we have adapted disciplinary tools and concepts for the context of our Climate Change Leadership Curriculum to the following:

1. Climate: Explaining how life is influenced by the environment and describing physical and chemical principles that affect the Earth's systems.
2. Energy: Describing how the flow of energy drives the cycling of matter.
3. Model and Maps: Interpreting and creating models and maps to understand (deep) time and scale of space
4. Earth Systems: Explaining the multitude of systems Earth is composed of as well as the system Earth belongs to.

2.3.1.4 Resources Guiding the Educator

In this section, we outline the resources available to educators in our curriculum. Ultimately, the resources would need to be accompanied by high-quality, context-specific professional development, and we outline our approach to PD in the following section on plans for implementation. By grounding the resources in clear standards, educators can readily identify the skills students need to develop and demonstrate throughout a climate change curriculum. To illustrate the product of this standards-based curriculum design, we have created an example of a 12-week curriculum resulting in a final student project. Anchor skills, process-based skills, and disciplinary tools and concepts are all integrated into each unit within the Climate Change Leadership Curriculum, emphasizing the importance of inquiry and experiential learning. As a result, each unit is based on at least one standard from each sub-section of skills:

Unit 1: Becoming Inquirers.
Unit 2: What is Climate? Why Does Climate Change?
Unit 3: Challenges to Understanding Climate Change.
Unit 4: Defining Problems to Solve in the Context of Climate Change.
Unit 5: Cognitive Biases and Logical Fallacies in Climate Change.
Unit 6: Gathering Evidence from Credible Sources.
Unit 7: Brainstorming Strategies to Address Climate Change.
Unit 8: Getting Feedback on Climate Change Strategies.
Unit 9: Creating a Criteria for Effective Climate Change Solutions.
Unit 10: Confronting Tradeoffs in Climate Change Policies.
Unit 11: Telling Your Story About Climate Change.
Unit 12: Taking Action in Addressing Climate Change.

In each unit, we provide a series of activities that guide teachers on how students could reach the standards. These activities are organized based on which essential question they address in the lesson so that teachers can understand the purpose whilst engaging in the activity and thus, modify it to meet the needs of their classroom. In Appendix A, we provide an overview of our activities.

The pedagogy incorporated in the activities is designed to reflect the skills and inclinations associated with the learning objectives. The teacher exposes students to ideas and sources and facilitates their ability to make meaning through individual and group analysis and reflection. If the teacher were to simply present the material, draw conclusions, and assign projects for the students, it would actually undermine the message that students should have agency in the process of learning and civic action. In fact, it would implicitly suggest that the voices of students do not matter. With that established, the teacher has an essential role to play when it comes to ensuring that there is a sense of accountability to the deeper mission of learning. The clarification of this mission is important to highlight the role of the teacher in terms of ensuring that the learning space is maintained and that students follow through with their work. While methods for ensuring accountability can vary, most methods generally include standard-based rubrics that are used to evaluate student work.

This is not to suggest that teachers cannot present information to students. On the contrary, teachers should use their understanding to help ensure that students can access information in a clear and engaging way. At times, the teacher may be uniquely suited to directly impart information, and oftentimes the teacher plays a role in curating information. But critical thinking on the part of the students should be encouraged at all times as students "decode" the messages of the teacher, and the teacher should model the skills and habits of inquiry and curiosity as an essential means of transferring those skills and habits to students. Furthermore, in the context of an issue as complex as climate change, there is a good chance that a teacher would be much more credible about some aspects of the issue than others, and as students pursue their own projects, the teacher plays the role of a coach and guide in the process. As the course progresses, students are provided opportunities to play these roles as well, supporting each other in the learning process as they develop skills of collaboration.

Lastly, we provide a student resource that serves as a project template, guiding the students through the process-based skills – essentially the process of creatively solving problems and taking informed action. This project template also serves as a resource for the educator as it provides an opportunity for students to practice the skills touched upon in the units through formative assessments related to a topic that they had agency and choice over (See Appendix B for an overview of the student project). Furthermore, it can be used as a summative assessment in the form of a meaningful capstone project. As a result, not only does it inform teaching throughout the curriculum, but it also allows for meaningful assessment of the process so that students can internalize the process of creative problem-solving and transfer it to other contexts throughout their life.

2.4 Implementation and Program Theory

There are several aspects of the context for delivery that are essential for the curriculum to make an impact, of which the teacher is the most important. In particular, the teacher interested in implementing the curriculum would need to have the skills to deliver the lessons as well as an understanding of the content of each lesson. Together with other faculty and school leaders, they would also need to be able to assess contextual factors that would dictate the ways in which the curriculum could be adapted to align effectively with the particular priorities and constraints of the institution.

The current implementation plan involves seeking feedback from teachers to learn what sort of professional development they might need and how they might adapt the curriculum to their contexts. As we learn from teachers about those needs, we can work with them to create modified versions of the curriculum that would be suited for implementation under different circumstances with consideration for various constraints in terms of time and/or other resources. Ultimately, we will have a wider variety of options for both professional development and implementation that teachers and school leaders could review. With that established, it would still be important to further align both the professional development and the curriculum with the needs of a particular school or course.

In order to find schools and teachers who would be interested in implementing the curriculum and giving us feedback, we have been working closely with the Arava Institute to identify alumni who may be involved in secondary education. Through this process, we were introduced to Fareed Mahameed. Fareed works with the Hura Agricultural Sustainable Farm, a program in the Negev Desert designed to serve as a center for research, education, and training for primary, middle and high schools in the region with a focus on integrating experiential education to learn about nature and the environment. We spoke with Fareed about the Climate Change Leadership Curriculum, and he went on to establish a partnership between the Hura Agricultural Sustainable Farm and the Al Taqwa Secondary school in Rahat, Israel, where he would teach the curriculum in collaboration with high school teacher Samah Matany. With the partnership, we worked together to adapt the curriculum to meet the needs and constraints of the Al Taqwa school. We also scheduled regular times to speak with Fareed to offer support in his delivery of the curriculum and to solicit feedback regarding the experiences of the different stakeholders involved.

In addition to conversations with the teacher implementing the curriculum, we will also collect student work, assessments, and surveys in order to gain a more complete picture of the experiences and perspectives necessary to inform the way forward. Furthermore, surveys with questions linked to each of the standards for the curriculum will be administered at the start and end of the course to gain insight into the efficacy of the curriculum. In Al Taqwa, Fareed administered a survey to the students that tested basic knowledge and confidence about climate change linked to the disciplinary tools and concepts, process-based, and anchor standards. In addition, we had survey questions regarding their general attitude about climate change leadership. The survey results provide a baseline for us to be able to measure impact through a post-survey. It also provided insight into the population of students we were working with. For example, Fareed noted that most students did not believe that individual actions could lead to climate change mitigation, which allowed him to consider how he could use specific units in systems thinking to address this attitude.

As students engage in the course, they will also have opportunities to demonstrate their assimilation of ideas related to climate change and the process of policy analysis through collaborative projects, personal reflections, and written assessments. Student project proposals that extend beyond the contained structure of the class will be the key indicator of their ability to transfer the understandings acquired during the course to new contexts. Finally, feedback from alumni will be solicited to measure their motivation and ability to continue to engage with the issue of climate change.

2.5 Conclusion

It is not sufficient for climate change education to focus on knowledge acquisition, and, even when students have opportunities to apply knowledge to address real-world problems related to climate change, an essential objective of climate change education is still left unmet. In particular, students need to understand the complexities of each step of the processes associated with thoughtful civic engagement in order to be enabled to truly lead in mitigating climate change. As students develop their understanding of how to affect change within complex systems, the standards and assessments must be aligned to these process-oriented skills in addition to the content knowledge. Ultimately, as students realize their potential to integrate thinking and practice, they will be able to affect meaningful change.

While a curriculum can provide the pedagogical frameworks and lessons to foster climate change leaders, a curriculum alone does not lead to successful implementation. Change comes about when the context-specific needs and interests of diverse stakeholders are considered and training and support is customized for teachers involved in the delivery of the curriculum. As we continue to engage in this work, there is potential to build an expanding network of educators and schools dedicated to fostering climate change leaders. By changing the way climate change is being taught, we hope that several high school students will be inquiring and observing the various impacts of climate change within their community and then be equipped, motivated, and empowered to take actions just like the ninth graders in Rahat, Israel.

Appendices

Appendix A: Climate Change Leadership Curriculum

This is a modified version of our curriculum with brief descriptions. The full curriculum can be accessed: https://climatechangeleadership.weebly.com/

Standards: Students will be able to….

Anchor Skills

A. Empathy

1. Communicate with people to understand why they do things, what their physical and emotional needs are, how they think about the world, and what is valuable to them.
2. Engage in constructive dialogue and public messaging connected to controversial issues, seeking to transform harmful dynamics and promote positive change.

B. Systems Thinking

1. Outline how actions can affect multiple stakeholders.
2. Identify causal webs of external and internal factors within and between systems.

C. Media Literacy

1. Critically analyze media in terms of origins, purpose, content, values and limitations.
2. Metacognitively reflect on personal biases and blind spots.

D. Collaboration

1. Work effectively with others through aligned goals.

Disciplinary Tools and Concepts

A. Climate

1. Explain how life, including human life, influences and is influenced by the environment
2. Describe physical and chemical principles that are unchanging and drive both gradual and rapid changes in the Earth system

B. Energy

1. Describe how the flow of energy drives the cycling of matter.

C. Models and Maps

1. Interpret and create models and maps to understand (deep) time and scale of space.

D. Earth Systems

1. Explain the multitude of systems Earth is composed of as well as the system

Step 1: Defining Problems

A. Develop compelling questions that promote inquiry around key disciplinary concepts and relevant, current issues.
B. Craft meaningful and actionable problem statements through the lens of the affected human population based on analysis and synthesis of diverse, relevant perspectives.

Step 2: Gathering Evidence

A. Determine the types of sources that will assist in understanding the problem and potential solutions.
B. Gather relevant information from various and diverse sources and evaluate for credibility.

Step 3: Generating Solutions

A. Synthesize evidence to generate a wide range of ideas to reach innovative solutions to a problem.
B. Create externalized prototypes to communicate potential solutions and receive feedback.

Step 4: Drawing Conclusions

A. Identify selection criteria and apply the criteria to different prototypes of potential solutions.
B. Synthesize the assessment of the relative strengths and weaknesses of different outcomes to draw conclusions about how to proceed.

Step 5: Communicating and Acting

A. Communicate the results of the process with relevant explanations and confront trade-offs.
B. Take direct action towards implementing the results of the analysis.

Unit 1: Becoming Inquirers

Essential Questions

- Why is inquiry important?
- What is inquiry and how does inquiry differ from advocacy?
- How do you develop skills required to engage in inquiry?
- How do you overcome common challenges when it comes to engaging in inquiry?
- What are the benefits of engaging in inquiry?

Rationale

In responding to the risks associated with climate change, it can be tempting to start with advocacy. We may want others to understand the dangers and feel inspired to join the effort to mitigate the risks. But this approach may overlook basic assumptions we bring to the table, thereby limiting our understanding of relevant perspectives that could inform our views on the goals and/or our understanding of the obstacles to progress. Unlike advocacy, inquiry starts from a place of curiosity. One way to think of the distinction between inquiry and advocacy is to consider different purposes for asking questions.

- **Advocacy** = Teach, Lead, Persuade.
- **Inquiry** = Learn, Understand, Explore.

Aligned Standards
Anchor.A.1 Empathy: Communicate with people to profile why they do things, what their physical and emotional needs are, how they think about the world, and what is valuable to them.
Anchor.A.2 Engage in constructive dialogue connected to controversial issues, seeking to transform harmful dynamics and promote positive change.
1.A Develop compelling questions that promote inquiry around key disciplinary concepts and embedded enduring issues.

Activities

1. Take a Stand on a Spectrum – Teacher presents a statement such as "The point of engaging in discourse with people who hold opposing views is to persuade others to change their views," while students line up on a spectrum from "strongly agree" to "strongly disagree." Students share their rationale, which might influence others to change their opinions, which allows students to understand how engaging in inquiry impacts conversations and outcomes.
2. Gallery Walk – Teacher prepares the following prompt:

   (a) "What are 'hot topics' that can lead to being blindsided by emotions?"
   (b) "What do you look, sound like, feel like when you're blindsided by emotion?"
   (c) "When you are blindsided by emotion, what are the impacts on the conversation?"
   (d) "What can you do when you feel like you may be blindsided by emotion? What can you do when you see others in this situation?"

   Students are divided into groups to write their own answers and rotate to observe other remarks and add additional comments.
3. Inquirer and Coach – Using the hot topics identified in the Gallery Walk Activity, teacher and a student will model a discussion while the student plays the role of an inquirer. Teacher takes a view in opposition to the student's firmly held beliefs. Other students can volunteer as coaches.
4. Take Space and Make Space – Teacher poses the question: "How likely are you to take space when you have an idea you want to communicate?" From class discussion, teacher establishes the purpose of inquiry and classroom norms.

Unit 2: What Is Climate? Why Does Climate Change?

Essential Questions

- What are compelling questions about climate change?
- What is a compelling question and what is a supporting question?
- What is climate and how does it relate to energy?
- What are long term vs. short term natural causes of climate change in Earth's history?

Rationale

According to the College, Career, and Civic Life (C3) Framework for Social Studies Standards, **compelling questions** "deal with curiosities about how things work; interpretations and applications of disciplinary concepts; and unresolved problems that require students to construct arguments in response." They are important, based on application of knowledge, and debatable with more than one possible answer.

Compelling questions are imperative to drive inquiry and curiosity before students engage civically. But students need **supporting questions** which provide the subject matter scaffolding needed before students can develop and engage in compelling questions on climate change. What is climate? Why does climate change? These are examples of important supporting questions for students to explore before looking into compelling questions such as "How can we take action to mitigate climate change?"

Standards

Anchor.D.1 Collaboration: Work effectively with others with goal alignment.
DTC.B.1 Energy: Describe how the flow of energy drives the cycling of matter.
DTC.D.1 Systems: Explain the multitude of systems Earth is composed of as well as the system Earth belongs to.
1.A Develop compelling questions that promote inquiry around key disciplinary concepts and embedded enduring issues

Activities

1. Learning Stations #1 – Teacher introduces and answers three supporting questions: what is climate? What is energy? How is energy related to climate (Teacher Friendly Guide, pg. 38). Students then go to learning stations to learn how energy flows through Earth systems and affects the climate. They will create and evaluate comprehension "tests" (supporting questions) for other students.

   (a) Air = atmosphere, (TFG, pg. 28–29, 30)
   (b) Water = hydrosphere, (TFG, pg. 28–29, 30)
   (c) Ice = cryosphere, (TFG, pg. 31–32)
   (d) Land = geosphere, (TFG, pg. 32–33)
   (e) Life = biosphere, (TFG, pg. 33)

2. Learning Station #2 – Teacher introduces a compelling question: "To what extent is climate change caused by human activities?" Students then go to learning stations to learn about the different natural causes of climate change. They will create essay questions (compelling questions) and then evaluate other students' works.
   (a) Plate tectonics (TFG, pg. 41–42, pg. 48)
   (b) Earth's orbit around the sun, Milankovitch cycles (TFG, pg. 42, 44)
   (c) Heinrich events (TFG, pg. 42)
   (d) Younger Dryas, Little Ice Age, El Nino (TFG, pg. 43)
   (e) Sun (solar flares, sunspots) (TFG, pg. 43)

Unit 3: Challenges to Understanding Climate Change

Essential Questions

- Why do we need systems thinking when addressing climate change?
- What are systems?
- How does systems thinking apply to climate change?
- How does human behavior influence climate change?

Rationale

Why is climate change so difficult to address? Understanding climate change involves learning about the complex interactions between multiple systems across disciplines. This is why it is imperative for students to develop skills in **systems thinking**, which is "a way of helping a person to view systems from a broad perspective that includes seeing overall structures, patterns and cycles in systems, rather than seeing only specific events in the system." As students develop their understanding of the ways interconnections between the elements of a system determine the behavior of the system, they can more effectively identify levers for systemic change. With systems thinking, students can turn their inquiry into some of the primary steps for taking informed action, identifying problems they might want to tackle when addressing climate change.

Standards

Anchor.B.1 Systems Thinking – Identify causal webs of external and internal factors.

0.A.1 Explain how life, including human life, influences and is influenced by the environment.

0.C.1 Interpret and create models and maps to understand (deep) time and scale of space.

1.B Craft meaningful and actionable problem statements through the lense of the affected human population based on analysis and synthesis of diverse relevant perspectives.

Activities

1. Create a System – After showing the concept of a system (https://www.youtube.com/watch?v=mNelPDgOcjs&list=PLfhKSsmO71c0V50JIR7io_

SEOFdGuyNS3&index=2, Climate Interactive, "The Whole System"), teacher introduces the carbon cycle system (TFG, pg. 47). Each group of students chooses one part of the carbon cycle system and maps out how it relates to other parts of the systems as well as outside of the system.

2. Iceberg Diagram – Teacher introduces the iceberg diagram (https://www.youtube.com/watch?v=MIIQ8Wfwu7c, Climate Interactive, "Iceberg Thinking Part 1"). Students read about the common causes of climate change (TFG, pg. 99–103). After, in small groups, students will choose an environmental issue from current events and ask "Why?" 5 times to create an iceberg diagram. Teacher asks students to identify **areas of high leverage-** the parts of a system that, when adjusted, will produce a significant change in the whole system.
3. Causal Diagrams – Teacher introduces causal diagrams through a video (https://www.youtube.com/watch?v=UgZTXf5PDis&list=PLfhKSsmO71c0V50JIR7io_SEOFdGuyNS3&index=6&t=0s, Climate Interactive, "Causal Diagrams"). In small groups, students refer back to their iceberg diagram and brainstorm one solution. Then, they create causal diagrams from their solution to summarize the effects of their solution.

Unit 4: Defining Problems to Solve in the Context of Climate Change

Essential Questions

- What are meaningful and actionable problem statements in the context of climate change?
- How is human life influenced by climate change?
- How do we identify root cause problems and areas of high leverage in climate change?
- How do we empathize to create actionable and meaningful problem statements?

Rationale

Defining a meaningful problem is the first step in policy analysis, design thinking, and conflict transformation. When understanding a problem, there's always a risk of jumping to a solution too quickly. Personal biases and blind spots about the causes of the problems can lead to false conclusions. That is precisely why it was so important to cultivate curiosity and build the skills of inquiry first.

There are also other challenges associated when creating problem statements specifically in the context of climate change due to psychological biases or political rhetoric. Ultimately, by applying the process of inquiry to assess real world problems associated with climate change in their region, students can begin to develop a collaborative project of their choice.

Standards

Anchor.A.1 Empathy: Communicate with people to understand why they do things, what their physical and emotional needs are, how they think about the world, and what is valuable to them.

Anchar.C.2 Media Literacy: Metacognitively reflect on personal biases and blind spots.
0.A.1 1. Explain how life, including human life, influences and is influenced by the environment
1.B Craft meaningful and actionable problem statements through the lense of the affected human population based on analysis and synthesis of diverse, relevant perspectives

Activities

1. Carousel – In small groups, students find local problems related to climate change. Students map out the root causes using the iceberg diagram. Students rotate and debrief on what are common effects of climate change on everyday life.
2. Empathy Interviews – Using a common effect from the carousel activity, students identify an affected stakeholder. Teacher models how to conduct an empathy interview while students practice interviewing each other:
   (a) Jobs and Tasks: What is this stakeholder trying to get done in their work or in their life?
   (b) Pains: What annoys this stakeholder or prevents them from getting their jobs done?
   (c) Gains: What are the outcomes or benefits that your stakeholder wants?
   (d) For example: Jordanian residents are trying to use water for basic drinking needs, but water is too expensive. They want cheap and easy ways to access water.
3. Climate Change Leadership Project – Students begin "Step 1: Defining Problem" of their own project. While students identify stakeholders to interview, teacher should guide students:
   (a) "Who would you specifically be interested to speak with regarding your topic?"
   (b) "Write why you would be interested to speak with them, how they could inform your understanding of the problem, and how you assess their credibility on the specific topic?"
   (c) "Compile questions that you would be interested to ask"

Unit 5 Cognitive Biases and Logical Fallacies in Climate Change

Essential Questions

- How do I limit the potential for cognitive biases and logical fallacies to influence my understanding of climate change?
- What are my own cognitive biases in the context of climate change?
- What are common logical fallacies in the context of climate change?
- How do I interpret and create models and maps to understand the scale of time and space in climate change?

Rationale

With a defined problem statement, the next step is to gather evidence to understand the problem more holistically and to inform different ways of addressing the problem.

Before diving into research, it is important to understand your own cognitive biases and logical fallacies. Students should therefore learn what factors influence how they think to understand their knowledge gaps. A common bias is the **availability heuristic**, which is when we rely on immediate examples rather than information that is grounded in extensive data. Given the scale of time and space in climate change, students will learn how to interpret models and maps as one example of how to combat a cognitive bias.

Standards

Anchor.C.2 Media Literacy: Metacognitively reflect on personal biases and blind spots.

0.A.1 Explain how life, including human life, influences and is influenced by the environment

0.C.1 Interpret and create models and maps to understand (deep) time and scale of space.

1.A Determine the types of sources that will assist in understanding the problem and potential solutions.

Activities

1. Illustrating Biases – After reading about the "Factors that Influence How We Think" and cognitive biases in TFG (pg. 218–219), students journal about their own biases and blindspots that could undermine their understanding of climate change. The teacher then facilitates a discussion of strategies to minimize the impact of cognitive biases, including:

   (a) Seeking feedback from diverse perspectives
   (b) Separating ideas from individuals.
   (c) Trying to focus less on existing processes, strategies, and technologies and envision new processes, strategies, and technologies.
   (d) Being aware of your cognitive biases.

2. Fallacies – Teacher introduces common types of logical fallacies (https://d32ogoqmya1dw8.cloudfront.net/files/integrate/teaching_materials/change_inthe_air/what_logical_fallacies_pdf.v5.pdf). Student uses the internet to browse news articles or videos related to climate change and identify the specific logical fallacies.
3. Scales and Maps – Teacher shows how models that display patterns over time can help the understanding of climate change (https://www.bloomberg.com/graphics/2015-whats-warming-the-world/). Students work in groups to identify different data points on an assigned climate change effect across time in any region to draw conclusions about the effects of climate change. Assigned effects can come from TFG (pg. 198–208): rising sea levels, heat waves, heavy rainfall, drought, extreme weather, reduced snowfall and snowpack.

Unit 6: Gathering Evidence from Credible Sources

Essential Questions

- What is reliable evidence I can use to allow me to understand my problem and potential solutions in addressing climate change?
- What are the values and limitations of sources?
- How do I use qualitative data as evidence?
- What are existing mitigation strategies to address climate change?

Rationale

Before thinking of possible solutions to the problem statement students have chosen, they need to conduct research. This is a valuable skill, but with the technology and internet students have access to a wealth of information. As such, they need to develop the skills to navigate the resources at their disposal. Now, **media literacy**, which includes the ability to access, analyze, and evaluate sources is increasingly important.

Conducting research is hard, but students will need to learn this valuable skill in order to understand the problem better and to look at common solutions. Oftentimes, students do not have to reinvent the wheel, but rather evaluate the extent to which common solutions would work in their chosen context. Teaching students how to conduct interviews to obtain this information will not only excite students, but will also allow them to research context-driven solutions while cultivating an open mind to new and innovative solutions as well!

Standards

Anchor.C.2 Media Literacy: Critically analyze media in terms of origins, purpose, content, values, and limitations.
0.A.1 Explain how life, including human life, influences and is influenced by the environment
0.C.1 Interpret and create models and maps to understand (deep) time and scale of space.
1.A Gather relevant information from various and diverse sources and evaluate for credibility.

Activities

1. Values and Limitations of Sources – Teacher models how to annotate sources to analyze their credibility through a reading from Project Look Sharp's: Discourse or Disinformation Lesson Plan" (pg. 75–78).

   (a) Origin: who, what, when, where publishing info, key details about origin.
   (b) Purpose: Goal, informative/argumentative, intended audience
   (c) Content: what is the summary of the source?

   From the origin, purpose, and content, teacher discusses with the students what might be the value or limitation of the source.

2. Collecting Data – Each group of students is assigned a mitigation strategy (TFG, pg. 152–175). They will produce a poster that summarizes the strategy as well as the pros and cons in the local context.
   (a) Renewable Energy
   (b) Nuclear Energy
   (c) Energy Efficiency and Conservation
   (d) Carbon Capture and Storage
   (e) Land Use: Forests, Soils, and Agriculture
   (f) Waste Management
3. Climate Change Leadership Project – Students begin "Step 2: Gathering Evidence" of their own project. While students identify stakeholders they want to interview, teachers should guide students in the similar fashion in Step 1:
   - "Who would you specifically be interested to speak with regarding your topic?"
   - "Write why you would be interested to speak with them, how they could inform your understanding of the problem, and how you assess their credibility on the specific topic"
   - "Compile questions that you would be interested to ask"

Unit 7: Brainstorming Strategies to Address Climate Change

Essential Questions

- Based on my evidence, how do I creatively generate solutions to address climate change?
- What are mitigation and adaptation strategies to address climate change?
- How do I synthesize my evidence to develop relevant insights?
- How do I spark innovative ideas to address climate change?

Rationale

In addressing climate change, it is possible to analyze options in terms of mitigation and adaptation. **Mitigation** would involve using current technologies to address climate change by seeking to reverse the rate of increasing average global surface temperature (e.g. seeking to replace fossil fuels with renewable energy). In contrast, **adaptation** is focused on ways to minimize the impact of climate change that is happening and will continue into the future (e.g. new construction according to building codes designed to withstand flooding or other extreme weather related events that are likely to increase in frequency). **Innovation** is about developing new ways to address climate change with a focus on developing technologies and strategies that do not yet exist. Now that students have gathered background evidence, they will generate a broad range of ideas that can include different combinations of mitigation, adaptation, and innovation before using relevant criteria to narrow their choices in their context.

Standards

Anchor.D.1 Collaboration: Work effectively with others through aligned goals.
0.A.1 Explain how life, including human life, influences and is influenced by the environment
3.A Synthesize evidence to generate a wide range of ideas to reach innovative solutions to a problem.

Activities

1. Adapting to Climate Hazards – Teacher presents adaptation strategies: relocate and retreat, adjust infrastructure, renew and conserve natural systems, make land use changes, modify management and operations, diversify to increase resilience, social innovation, risk management, and policy changes. Using TGF (pg. 198–208), students work in groups to come up with adaptation strategies for one of the described climate hazards. Students rotate and add different ideas for adaptation strategies to different posters.
2. Affinity Diagramming – Teacher presents a brainstorming strategy called affinity diagramming (https://www.youtube.com/watch?v=UynxDyr0lAo). Using the posters from the "Adapting to Climate Hazards" activity, students create a post-it note for each adaptation strategy. In groups, students group the post-its into themes and put headings onto these groups.
3. Attribute Listing – Teacher presents a strategy called attribute listening, where column headings are attributes.

	Strategy	Incentive	Stakeholder	Medium
Subsidize solar panels	Mitigation	Giving rewards	Households	Policy
Taxing businesses based on carbon consumption (carbon taxes)	Mitigation	Punishing bad behavior	Businesses	Policy
Influence consumer water conservation behavior with the use of smart meters and pricing	Adaptation	Increasing awareness	Consumers	App / Technology
Foster community networks that find and help at-risk populations	Adaptation	Social Pressure	Communities	Social Networks

Teacher states that one strategy is to pick random attributes (such as the highlighted boxes in the table) and come up with new ideas. Students can also come up with new strategies, incentives, stakeholders, or medium.

4. Climate Change Leadership Project – Students begin "Step 3: Generating Solutions."

Unit 8: Getting Feedback on Climate Change Strategies

Essential Questions

- How do I receive honest feedback from local stakeholders on potential ways to address climate change?
- How do I create prototypes to communicate potential solutions?
- How do I test a hypothesis to receive meaningful and usable feedback from local stakeholders?
- What is geoengineering and how would local stakeholders be affected by this strategy?

Rationale

While students explore various solutions, it is better for them to engage with local stakeholders as a planning and implementation strategy for the solution they choose in addressing climate change. First of all, according to **lean design thinking** models, the earlier you get feedback, the less costly it is in both time and resources. Students should receive feedback on their potential solutions before they build on those solutions in the next step in which they draw conclusions with regard to the best path forward. Second of all, when local stakeholders are engaged, it creates possibilities to generate more targeted solutions and opens opportunities for stakeholders to be involved in the implementation of those solutions. Students will explore how to communicate with local stakeholders through creating simple prototypes. We will model how to do this while learning an innovative climate approach to addressing change- **geoengineering**.

Standards

Anchor.A.1 Empathy – Communicate with people to understand why they do things, what their physical and emotional needs are, how they think about the world, and what is valuable to them.

0.C.1 Interpret and create models and maps to understand (deep) time and scale of space.

0.D.1 Explain the multitude of systems Earth is composed of as well as the system Earth belongs to.

3.B Create externalized prototypes to communicate potential solutions and receive feedback.

Activities

1. Prototyping Pt. 1 – Teacher introduces geoengineering: "large-scale technological effort to change Earth's climate by removing carbon dioxide from the

atmosphere or decreasing heat received from the sun." Teacher points out that it is easier to give feedback about a complicated solution through a visual. In small groups, students are assigned a geoengineering technique from TFG (pg. 182–189): enhanced chemical weathering, marine cloud brightening, stratospheric aerosol distribution, and surface albedo alteration. They must create a visual prototype of the process, present to the class, and receive feedback (pros and cons) of using the technique.

2. Prototyping Pt. 2 – Teacher presents a visual of ocean fertilization, a geoengineering solution (https://marinefoodchainsmn.tumblr.com/post/73374299312/solution-iron-fertilization). Students are divided into small groups and are assigned a stakeholder: politician, scientist, iron sulphate producers, fishermen and fisheries, community citizens (who use water). Students will come up with a list of questions to ask the stakeholder if they want to get feedback on ocean fertilization with the visual prototype.
3. Climate Change Leadership Project – Students continue "Step 3: Generating Solutions" of their own project.

Unit 9: Creating Criteria for Effective Climate Change Solutions

Essential Questions

- How do I create selection criteria for effective climate change solutions to the problem I am trying to address?
- What are common selection criteria to evaluate climate change solutions?
- How do I apply a cost-benefit analysis to climate change solutions?
- What are several systems and stakeholders to consider when evaluating climate change solutions?

Rationale

It is often very tempting to immediately choose a solution, but it is important to first systematically compare the different options. One of the strategies for students to use before drawing solutions for the best path forward is to identify **selection criteria** which are standards against which different solutions would be measured to see how they meet the goal of solving the problem. Students will work together to identify and prioritize selection criteria before narrowing down their solutions. Common selection criteria that we will explore include effectiveness and efficiency (cost-benefit analysis).

Standards

Anchor.B.1 Systems Thinking – Outline how actions can affect multiple stakeholders.

0.A.1 Explain how life, including human life, influences and is influenced by the environment.

4.B Identify selection criteria and apply the criteria to different prototypes of potential solutions

Activities

1. Four Corners – Teachers will state an example of a mitigation, adaptation, and innovation strategy. Students will physically move to a corner in the classroom labelled as "Strongly Agree," "Agree," "Strongly Disagree," or "Disagree," reflecting their opinion of how effective the strategy is. Class discusses why they made their decision, leading them to develop selection criteria such as effectiveness, convenience, time, efficiency, sustainability, equity, feasibility, etc.
2. Cost-Benefit Analysis – Teacher explains the concepts of costs and benefits as well as strategies to use when conducting the analysis:

   (a) Bottom up = one constructs detailed inventories of systems and their components in order to explore the relevant costs of specific adaptation methods.
   (b) Top down = one uses aggregate data on a system and makes assumptions about the additional costs needed to apply adaptation techniques.

   Students practice cost-benefit analysis from Box 9.2 Classroom exercise on cost-benefit assessment (TFG, pg. 198).

3. World Climate Simulation – "The World Climate Solution is a role playing exercise for groups in which participants engage in the UN climate change negotiations. It is unique in that it uses an interactive computer model to rapidly analyze the results of the mock-negotiations during the event." All directions, materials, and the program are available on this website: https://www.climateinteractive.org/programs/world-climate/
4. Climate Change Leadership Project – Students continue "Step 4: Drawing Conclusions" of their own project.

Unit 10: Confronting Tradeoffs

Essential Questions

- How do I systematically draw conclusions with regard to the best proposed solution to address climate change?
- How do I prioritize selection criteria?
- How do I project outcomes in the context of a proposed solution?
- How do I apply selection criteria to projected outcomes in order to assess the strengths and weaknesses of proposed solutions?

Rationale

The first instinct is to compare the different proposed solutions based on the selection criteria developed in the previous lesson. However, in order to evaluate solutions, we should be focusing on what the solutions would lead to. In other words, we need to look at the **projected outcomes**. Once the selection criteria are defined and the projected outcomes of alternative courses of action have been established, it is necessary to apply the criteria to the outcomes. With this information, it will be possible to compare the options systematically – an essential step towards deciding between them.

Standards

Anchor.B.1 Systems Thinking: Outline how actions can affect multiple stakeholders.

Anchor.B.2 Systems Thinking: Identify causal webs of external and internal factors within and between systems.

DTC.D.1 Earth Systems: Explain the multitude of systems Earth is composed of as well as the system Earth belongs to.

5.B Synthesize the assessment of the relative strengths and weaknesses of different outcomes to draw conclusions about how to proceed.

Activities

1. Values-based Reasoning – In small groups, students must identify whether a strategy that limits the availability of water in a region would be an effective strategy in addressing climate change. Teacher then facilitates a discussion of how values-based reasoning is essential to suggest that other than looking at criteria such as efficiency and effectiveness, what concerns of equity could arise from this type of solution? How could students bring in valued-based reasoning in a way that doesn't open the possibility of simply affirming a worldview because it is prevalent in society?
2. Modeling Projected Outcomes – Using Eugene Bardach's *A Practical Guide for Policy Analysis*, teacher explains what an outcomes matrix is (pg. 46–65). Teacher asks students to choose 3 policy alternatives to reduce carbon emissions and create an outcomes matrix in the similar fashion:

	Projected Outcome on Selection Criteria #1	Projected Outcome on Selection Criteria #2	Projected Outcome on Selection Criteria #3
#1			

3. Climate Change Leadership Project – Students continue "Step 4: Drawing Conclusions" of their own project. Using Climate Generation, "Next Generation Climate for Grades 6–8" (https://curriculum.climategen.org/2017/NGC/NGC2017-5-4-Worksheet-Claim-Evidence-Reasoning.pdf) teacher models how to create a claim, evidence, and reason from the outcomes matrix. Example:

   (a) Claim: Requiring zero-energy capable homes is the best option for minimizing the impacts of climate change as a mitigation strategy.
   (b) Evidence: The estimated cost to implement a plan would be around $132 to $250 per ton of carbon dioxide abated.
   (c) Reasoning:

      (i) Why does your evidence support your claim?

1. Ex: This cost effectiveness was calculated based on….

(ii) Why is your evidence important? In other words, why is this selection criteria prioritized?

1. Ex: Cost effectiveness is important given the limited government budget of….

Unit 11: Telling Your Story

Essential Questions

- What are effective ways to communicate the results of this process to garner support for a potential solution for addressing climate change?
- How do I talk about controversial issues and complex systems?
- How do I develop empathy for various stakeholders?
- What are key questions to ask when creating media messages

Rationale

If someone were to ask about the problem being addressed and the solution you propose, how would you answer? It depends on the audience. Imagine someone like your grandmother.

Imagine other important target audiences. How would you advocate effectively for the course of action you propose? How might you engage with somebody who does not believe climate change exists? This analysis can lead to insight into overlooked stakeholders and/or interests that may be important to take into consideration. In general, it is important to think through everyone who would be involved in successful implementation in order to ensure that it is as effective as possible.

Standards

Anchor.A.2 Empathy – Engage in constructive dialogue connected to controversial issues, seeking to transform harmful dynamics and promote positive change.
Anchor.D.1 Collaboration – Work effectively with others through aligned goals.
DTC.B.1 Systems Thinking – Outline how actions can affect multiple stakeholders.
5.A Communicate the results of the process with relevant explanations and confront trade-offs.

Activities

1. Debate Role-Play – Teacher presents “Rules of Thumb for Teaching Controversial Issues” (TFG, pg. 214–215). Students then watch a debate between Bill Nye the Science Guy and Nick Loris of the Heritage Foundation regarding climate change, taking notes of effective and ineffective techniques of discourse. In small groups, students re-enact the debate with the new insights from class discussion.

2. Teaching Others – Task students to teach another class the strategy they are proposing to address climate change. Before doing so, students should create an "Identity Chart" of the class they will be teaching for (Source: Facing History, https://www.facinghistory.org/resource-library/teaching-strategies/identity-charts)
3. Creating Media – Teacher introduces different ways of using language and perspective in teaching climate change (TFG, pg. 223–230). Students will create a post on social media with a graphic illustration to communicate their proposal using the techniques they have just learned. Before creating their media message, they must answer "Key Questions to Ask When Creating Media Messages" from Project Look Sharp.
4. Climate Change Leadership Project – Students continue "Step 5: Communicating and Acting."

Unit 12: Taking Action in Addressing Climate Change

Essential Questions

- What direct actions can you take towards implementing the results of your analysis beyond telling your story?
- How do I create SMART goals?
- How do I make a project timeline?
- How do I take the first step?

Rationale

Once the course of action is selected and the underlying story/rationale can be conveyed clearly and powerfully, students are ready to plan their own individual action. Sometimes this involves asking about ways to support an ongoing local initiative. It can also take the form of a new initiative in the school or in the broader community. Their next step is to break it down into goals that are Specific, Measurable, Achievable, Relevant and Time-based (SMART goals). We hope that the students will be motivated and enabled to become effective climate change leaders within their communities!

Standards

Anchor.A.2 Empathy – Engage in constructive dialogue connected to controversial issues, seeking to transform harmful dynamics and promote positive change.
Anchor.D.1 Collaboration – Work effectively with others through aligned goals.
DTC.B.1 Systems Thinking – Outline how actions can affect multiple stakeholders.
5.B Take direct action towards implementing the results of the analysis.

Activities

1. SMART Goals – Using MindTools, "SMART Goals," teacher shows an example of a SMART goal and annotates it. In their small groups, students attempt to create a SMART goal for their proposed solution. Students will

rotate and annotate the SMART characteristics in another group's SMART Goal.
2. Planning Projects – Using the Climate Change Connection's Youth Guide for Action, teacher walks through a project plan (https://climatechangeconnection.org/wp-content/uploads/2014/09/Climate_Guide_to_Action_en-1.pdf). After watching a video about Boyan Slat's Ocean Clean-up, students can practice outlining a project plan for Boyan.
3. Climate Change Leadership Project – Students finish "Step 5: Communicating and Acting." Teachers should have an assessment for students to present the culmination of their work to an authentic audience. Teachers should work with the school to determine how students can take the first steps in implementing their ideas to shift the learning from analysis to experience.

Appendix B: Climate Change Leadership Project – Student Version

This is a modified version of the student project guide. The full guide can be accessed: https://climatechangeleadership.weebly.com/

Step 1: Defining Problems

1. What is the problem I am trying to solve?
2. Root cause analysis: Why does this problem happen? Why? Why? Why? Why?
3. Stakeholder analysis: Who are the key stakeholders in the context of the chosen topic?
    (a) Who is directly impacted?
    (b) Who are local experts?
    (c) Who is already seeking to affect change?
    (d) Who else might be capable of affecting change?
4. Stakeholder profile map:
    (a) Jobs and Tasks: What is this stakeholder trying to get done in their work or in their life?
    (b) Pains: What annoys this stakeholder or prevents them from getting their jobs done?
    (c) Gains: What are the outcomes or benefits that your stakeholder wants?
    (d) Purpose, Credibility, Questions to ask
5. What is your problem statement now?

Step 2: Gathering Evidence

1. Record your sources using the following template

Origin	Purpose	Values and Limitations
Who, What, When, Where Publishing Info: Key Details:	Goal: Informative/Argumentative: Intended Audience:	

2. Plan your interview accordingly:
   (a) Purpose of interview
   (b) Information I know about the interviewee
   (c) Possible Questions that are open ended, build on background information, draw out interesting ideas and information, and cover a range of relevant topics (adapted from The Critical Thinking Consortium).

Step 3: Generating Solutions

1. Synthesize your evidence by doing an affinity mapping exercise.
2. Individually brainstorm solutions.
3. Prototype your solution. Directions are adapted from Stanford d.school, "See Beyond: A Systems + Design Investigation" https://dschool.stanford.edu/resources/seebeyond.
   (a) What is the solution? How does it work? How does it solve your problem?
   (b) Can you draw/visualize it here?
   (c) Who/How to Implement? How does this solution get created and into the system? Who are local stakeholders?
4. Interview Stakeholders
   (a) Identify the most relevant stakeholders.
   (b) What is your goal? What do you want to find out from them?
   (c) What are possible questions you might ask them for feedback.

Step 4: Drawing Conclusions

1. Selection criteria: What is your selection criteria? How important is this? Evidence
2. Outcomes Matrix
   (a) What is your baseline?
   (b) Create a matrix

	Criteria #1	Criteria #2	Criteria #3
Policy #1			
Policy #2			

3. Conclusion
   (a) Claim
   (b) Evidence (Data)
   (c) Reason
      (i) Why does your evidence support your claim?
      (ii) Why is your evidence important?
   (d) Repeat b and c.

Step 5: Communicating and Acting

1. Identity chart: referencing your previous stakeholder charts, which one is the most important to communicate your proposed solution to? Highlight important aspects of their identity.
2. Media message: Before you create your own appealing media message, answer the following questions about your stakeholders. Source: Adapted from Project Look Sharp
   (a) Authorship:
      (i) Who am I representing in making this?
      (ii) Who are my co-creators?
   (b) Purpose:
      (i) Why am I making this? Who is my target audience?
      (ii) What do I want people to do as a result of my message?
      (iii) What do I want people to think (or think about)?
   (c) Economics
      (i) Who might benefit from this message?
      (ii) Who might be harmed by it?
      (iii) Whose voices are represented or privileged? Omitted or silenced?
      (iv) What is my responsibility to my audience?
   (d) Content
      (i) What messages and impressions do I want to convey?
      (ii) What ideas, values, and information do I want to make explicit? Implied?
      (iii) What will I choose to leave out of this message, and why?
      (iv) Is my presentation of my information and ideas fair?
   (e) Context:
      (i) Where and how will I share this message with my audience?
      (ii) How might cultural context influence the way people interpret my message?

(f) Techniques:

  (i) What techniques will work best to communicate the messages for this audience, and why?
  (ii) Do I have (or need) permission to use this content?

(g) Interpretations:

  (i) How (and why) might different people interpret this differently?
  (ii) What do I learn about myself from my choices in making this?

(h) Responses:

  (i) How might people feel after hearing, reading, or viewing this message?
  (ii) What kinds of actions might people take in response to this?

3. Smart Goal:

  (a) **Specific**: Who is involved? What do I specifically want to accomplish? Where? For what purpose?
  (b) **Measurable**: How much? How many? What will I know when I have accomplished the goal?
  (c) **Achievable**: How will I accomplish this goal? What are the obstacles or constraints to overcome?
  (d) **Relevant**: How does the goal fit into the context of the area? How does it fit into your personal context?
  (e) **Time-based**: What I can do in a year? What can I do in six months? What can I do in 2 weeks? What can I do today?

4. Project Implementation Plan (adapted from Climate Change Connections)

  (a) List steps

Step Number	Steps Needed to Achieve Your Goal	How Will You Do This	What Help Will I Need	How will you know when you are successful?

  (b) What are possible challenges? How can you overcome this?

References

Al-Newashi, Q. (2002). *Towards improving the status of the formal and nonformal environmental education in Jordan* (Doctoral dissertation). Retrieved from Deutsche National Bibliothek. https://ediss.sub.uni-hamburg.de//volltexte/2002/696/

Andreou, N. (2020). Towards a generation of sustainability leaders: Eco-schools as a global green school movement for transformative education. In A. Gough, J. C.-K. Lee, & E. P. K. Tsang (Eds.), *Green schools globally: Stories of impact on education for sustainable development*. Springer. (in press).

Bardach, E. (2011). *A practical guide for policy analysis: The eightfold path to more effective problem solving*. Washington, DC: CQ Press.
Cairns, D. (2010). Investigating the relationship between instructional practices and science achievement in the inquiry-based learning. *International Journal of Science Education, 41*(15), 2113–2135. Retrieved from https://www.tandfonline.com/doi/full/10.1080/09500693.2019.1660927.
Climate Interactive. (2019). Retrieved from https://www.climateinteractive.org/
Cohen, S. (2014). *Understanding environmental policy*. New York: Columbia University Press.
Cyprus Ministry of Education and Culture. (2012). *SEMEP good practices first collection: Bridging cultures through science for sustainable environment*.
Eames, C., & Mardon, H. (2020). The enviroschools programme in Aotearoa New Zealand: Action-orientated, culturally responsive, holistic learning. In A. Gough, J. C.-K. Lee, & E. P. K. Tsang (Eds.), *Green schools globally: Stories of impact on education for sustainable development*. Springer. (in press).
EcoPeace Middle East. (2019). *Climate change, water security, and national security for Jordan, Palestine, and Israel*.
Gonzalez-Gaudiano, E., & Meira-Cartea, P. (2012). Climate change education and communication: A critical perspective on obstacles and resistances. In *Education and climate change* (1st ed., pp. 13–34). New York: Routledge.
Haber, T. L. (2019). *Environmental leadership seminar*. Israel: Arava Institute.
Intergovernmental Panel on Climate Change. (2018). *Special report on global warming of 1.5°C*. Retrieved from https://report.ipcc.ch/sr15/pdf/sr15_spm_final.pdf
Intergovernmental Panel on Climate Change. (2019a). *Climate change and land*. Retrieved from https://www.ipcc.ch/site/assets/uploads/2019/08/Edited-SPM_Approved_Microsite_FINAL.pdf
Intergovernmental Panel on Climate Change. (2019b). *The ocean and cryosphere in a changing climate*. Retrieved from https://report.ipcc.ch/srocc/pdf/SROCC_FinalDraft_FullReport.pdf
Karama, M. (2016). A comparative survey of environmental education goals between UNESCO framework and 10th grade Palestine curriculum. *International Journal of Curriculum and Instruction, 8*(2), 1–17. Retrieved from https://files.eric.ed.gov/fulltext/EJ1207324.pdf.
Ledley, T. S., Rooney-Varga, J., & Niepold, F. (2017). Addressing climate change through education. *Oxford Research Encyclopedia of Environmental Science*. https://doi.org/10.1093/acrefore/9780199389414.013.56.
National Council for Social Studies. (2017). *College, Career & Civic Life C3 Framework for Social Studies State Standards: Guidance for Enhancing the Rigor of K-12 Civics, Economics, Geography, and History*. Retrieved from https://www.socialstudies.org/sites/default/files/2017/Jun/c3-framework-for-social-studies-rev0617.pdf
Next Generation Science Standards. (2019, September 9). Retrieved from https://www.nextgenscience.org/
Platner, H. (n.d.). *An introduction to design thinking: Process guide. An introduction to design thinking: PROCESS GUIDE*. Stanford: Institute of Design at Stanford. Retrieved from https://s3-eu-west-1.amazonaws.com/ih-materials/uploads/Introduction-to-design-thinking.pdf.
Project Look Sharp: K-12 & Higher Ed. Media Literacy Lesson Plans: Ithaca College. (n.d.). Retrieved from https://www.projectlooksharp.org/
Reimers, F. (2020). *Educating global citizens*. Springer.
Sagy, G., & Tal, A. (2015). Greening the curriculum: Current trends in environmental education in Israel's public schools. *Israel Studies, 20*(1), 57–85. Retrieved from https://www.jstor.org/stable/10.2979/israelstudies.20.1.57?seq=1.
Selby, D., & Kagawa, F. (2013). *Climate change in the classroom: UNESCO course for secondary teachers on climate change education for sustainable development*. UNESCO.
Tal, T. (2020). Green schools in Israel: Multiple rationales and multiple action plans. In A. Gough, J. C.-K. Lee, & E. P. K. Tsang (Eds.), *Green schools globally: Stories of impact on education for sustainable development*. Springer. (in press).
UN CC: Learn. (2013). *Integrating climate change in education at primary and secondary level. Integrating climate change in education at primary and secondary level*. Retrieved from

https://www.uncclearn.org/sites/default/files/resource_guide_on_integrating_cc_in_education_primary_and_secondary_level.pdf

UNESCO & United Nations Environment Programme. (2011). *YouthXchange guidebook series: climate change and lifestyles*. Retrieved December 11, 2019 from https://unesdoc.unesco.org/ark:/48223/pf0000212876

UNESCO Associated Schools Network. (n.d.). Retrieved from https://aspnet.unesco.org/en-us/climate-education-education

UNESCO GAP Priority Action Areas. (2019, March 4). Retrieved October 3, 2019, from https://en.unesco.org/gap/priority-action-areas

UNESCO Institute of Statistics. (2020). Retrieved February 25, 2020, from http://uis.unesco.org/

United Nations. (2015). *Transforming our world: The 2030 agenda for sustainable development. Transforming our world: The 2030 agenda for sustainable development*. Retrieved from https://sustainabledevelopment.un.org/content/documents/21252030+Agenda+for+Sustainable+Development+web.pdf

United Nations Climate Change – Summit 2019. (n.d.). Retrieved October 3, 2019, from https://www.un.org/en/climatechange/education-key-climate-change.shtml

Zabel, I. H. H., Duggan-Haas, D., & Ross, R. M. (2017). *The teacher-friendly guide to climate change*. Ithaca: Paleontological Research Institution.

David Rhodes taught English in Nicaragua, and he taught history, philosophy, and English in Colombia. Then he began teaching social studies, Spanish, and service learning at the middle school level in Corning, New York. In addition to classroom instruction, David also facilitated leadership development programs, and he coordinated professional learning workshops related to civic education and media literacy. He is a graduate of the Education Policy and Management Masters program at the Harvard Graduate School of Education. David is passionate about fostering thoughtful civic leaders and creating opportunities for students to collaborate on meaningful projects as they engage in learning that transcends the walls of the classroom.

Margaret Wang was a secondary educator in social studies, economics, business management, and entrepreneurship in an international school in Bahrain. She has also taught English and Public Speaking in several regions in China and Japan. She is a graduate of the International Education Policy Masters program at the Harvard Graduate School of Education. She loves to empower students to realize their potential and become agents of social change.

Chapter 3
Creating a Culture of Shared Responsibility for Climate Action in Guatemala Through Education

Lina Lopez Lalinde and Carrie Maierhofer

3.1 Introduction

In Chap. 1 of this book, Fernando Reimers explains that climate change education (CCE) is one of the most critical avenues available to prepare individuals to understand the science behind changing climate trends and the impact of human action on the environment, as well as learn strategies for adapting to and mitigating the effects of climate change in their communities. This idea has been emphasized by leaders in international education and climate action, who have proclaimed that "education can bring about a fundamental shift in how we think, act, and discharge our responsibilities toward one another and the planet" (UNESCO 2017, p. 2). In Guatemala, a country particularly vulnerable to the effects of climate change, the groundwork has been laid for using education to combat these effects, both through the integration of environmental and climate topics into the country's Basic National Curriculum, as well as through proclamations by the government on the need to make climate change education a national priority. Unfortunately, these efforts have yet to create a culture of climate action and shared responsibility, according to the country's Minister of Education and its Minister of Environment and Natural Resources (MARN 2017). For this reason, it is essential to identify the gaps that have resulted in this failure. It is only through this examination that solutions can be crafted and implemented in order to meet the goal of creating a culture of shared responsibility and action against climate change in Guatemala.

To that end, this chapter first elaborates on Guatemala's vulnerabilities to the impacts of climate change and discusses educational efforts currently in place to equip communities with relevant climate adaptation and mitigation skills. Subsequently, it presents an overview of the literature identifying gaps in these

L. Lopez Lalinde (✉) · C. Maierhofer
Harvard Graduate School of Education (HGSE), Cambridge, MA, USA
e-mail: llopezlalinde@gse.harvard.edu

F. M. Reimers (ed.), *Education and Climate Change*, International Explorations in Outdoor and Environmental Education,
https://doi.org/10.1007/978-3-030-57927-2_3

efforts and describes an approach within the formal education system which works to address these gaps. More specifically, it will present the blueprint for a region-specific, school-centered guidebook that aims to explain how four particular components of a school's system can work together towards mitigating and adapting to climate change. The blueprint will also provide examples of the types of activity recommendations included in the guidebook, geared specifically towards helping a pilot school in Guatemala create a schoolwide culture of climate action.

3.2 Impact of Climate Change on Guatemala

As a country, Guatemala faces many concerning challenges, including an immense vulnerability to natural hazards, staggering numbers of people living in poverty, and malnourishment. These challenges will continue to worsen, as the climate continues to change through increased temperatures, extreme rainfall, droughts, floods, and other climate-related changes (USAID 2017). These climate changes pose a particular risk to the people of Guatemala, as they threaten to impact the country's agricultural sector, an industry which is both reliant on climate conditions and crucial for economic growth and stability. According to the Food and Agriculture (FAO) Policy Decision Analysis, "70 percent of [the country's] total land area [is] dedicated to agricultural and forestry activities" (FAO 2014, p. 1).

Guatemala has long been vulnerable to climate hazards. According to the Global Facility for Disaster Reduction and Recovery (GFDRR 2009), it is ranked fifth among countries with the highest economic risk to three or more hazards, particularly due to the vulnerability of its gross domestic product (GDP) to hazards. The GFDRR estimates that about 83.3% of the country's GDP is located within areas of risk. These areas are vulnerable to both low frequency-high impact events, including earthquakes, volcanoes, and hurricanes, and high frequency-low impact events, such as floods and landslides (GFDRR 2009).

Due to the great levels of inequality within Guatemala, a large part of the population actually lives in places that are not suitable for human habitation. In the Guatemala City metropolitan area, for instance, close to 600,000 people live in landslide-prone hillsides, while an additional 350,000 people live in plains located within flood zones on the Pacific coast and in the departments of Izabal and Petén (Alcarraz et al. 2012). According to the European Commission Humanitarian Aid Department's Disaster Preparedness Programme, a total of three million people live in high-risk areas within the country (Alcarraz et al. 2012). Making matters worse, these individuals often lack access to basic needs such as water and sanitation, as well as to public services such as education, health, security, and criminal justice. As a result of all these factors, it is in these places that the impact of climate change is the greatest (Hernández 2012).

Additionally, these climate hazards threaten to affect many key areas of individuals' lives, including children's schooling. Following the eruption of Volcano Pacaya in 2010, along with the subsequent tropical storm Agatha which brought tremendous rainfall, Guatemala suffered losses of around 640.3 million quetzales (USD 82.9 million) within the education sector alone, mainly as a result of damages to school buildings and infrastructure (CEPAL 2011). According to estimates from the United Nations Economic Commission for Latin America and the Caribbean (CEPAL 2011), about 6.9% of students enrolled within the country, which constitutes around 229,940 students, as well as 7810 teachers, were affected by these events.

Clearly, climate change poses a considerable risk to Guatemalans and, therefore, a great need exists for communities to become better informed about the impacts of climate change and better equipped with the skills to effectively mitigate these impacts. In 2017, the Guatemalan government stated that "education is one of the pillars of development of communities, with environmental education being particularly essential for the construction of sustainable and resilient societies, committed to the care, protection, and betterment of the environment" (MARN 2017, p. 16). In fact, the government has successfully integrated sustainable development and climate change topics in the Basic National Curriculum (MARN 2017). As a result, environmental education has become a permanent fixture in the national education system, and a commitment to adapting to and fighting against climate change using education as a vehicle for change has been established (CALAS 2010, Congress of Guatemala 2013).

Yet, despite the elevation of this topic to the policy agenda of the national government, the government has acknowledged that these efforts have been insufficient and have largely failed to create a culture of shared responsibility and climate action within society (MARN 2017). In a joint letter published within the 2017 Guatemalan National Policy on Environmental Education, the Ministers of Education and of Environment and Natural Resources stated that "every day, the population is more exposed to… the impact of climate change, a consequence of a social, economic and political system that fails to take into account our shared responsibility for the care and betterment of the environment" (MARN 2017, p. 7). The two ministries have also condemned society's inaction and particularly its lack of attention to the ways in which climate change is affecting the country's most marginalized groups (MARN 2017).

Building on the ministers' claims, we investigated the types of CCE programs currently being carried out within the country and found that the ministers' concerns are indeed reflected in large part within the Guatemalan education system. On one hand, the severe lack of CCE programming directed towards the country's most vulnerable populations, including Indigenous groups and its very large out-of-school population, evidences an indifference towards the experiences of those who have the most to lose from the country's inaction on climate change. Similarly, the overall lack of interdisciplinarity within the Basic National Curriculum, and of teacher professional development opportunities geared towards building competencies on environmental and climate topics, indicate a lack of acknowledgement among the country's educational authorities about the complex and interdisciplinary nature of this issue. The following section expands upon the gaps that have contributed to the inadequate sense of a climate action-oriented culture and to propose potential solutions to addressing these gaps.

3.3 The State of Climate Change Education in Guatemala and Opportunities for Improvement

There is currently ample will within Guatemala to find ways to leverage education as a means of empowering communities to protect themselves against the effects of climate change. The inclusion of CCE topics into the Basic National Curriculum, particularly, has made an understanding of climate change and its mitigation a priority and has created a space within which additional CCE efforts can be tested and implemented. The following section further explores the reasons why, despite this, a culture of shared responsibility and climate action has yet to be achieved. In particular, it presents a review of literature identifying gaps and failings in Guatemala's current CCE efforts and then presents what we believe to be the most appropriate solution for creating this culture of climate action within Guatemalan communities. Ultimately, we deemed the whole-school model to be the optimal approach for reaching this goal, given the scope of our project and our overall goal. In the sections that follow we hope that the reasons for our choice to move forward with the whole-school model will be made evident.

3.4 What Are the Major Gaps in Climate Change Education in Guatemala?

There are multiple sources of fragility in equipping the Guatemalan population with the knowledge and skills necessary to mitigate the impact of climate change. These include the lack of bilingual education and of adequate opportunities to educate the large Maya population, the considerable proportion of youth and adults who are out of school and who have achieved very limited formal education, and the lack of coherence and alignment among various components of the education system and a climate change curriculum.

3.4.1 Lack of Adequate Bilingual Education

One of the main concerns expressed by the Guatemalan administration about CCE was the lack of focus on the experiences of underrepresented and underprivileged groups in Guatemala. Indeed, the literature (Patrinos and Velez 2009; de la Garza 2016) confirms that the marginalization of these groups through inequalities within the education system continues to be an enormous challenge for the sector as a whole. One group of particular concern is the large number of students in the country's Indigenous schools, which experience the highest illiteracy rates in the country and generally perform the worst in key achievement indicators (Patrinos and Velez 2009). A 2009 study published by the Center for Economic and Social Rights for instance, revealed that literacy rates for rural, Indigenous young adults were at around 76%, compared to 96% for urban, non-Indigenous youth (2009). These data illustrate the clear necessity for institutions to devote significant resources to creating programs and policies aimed at meeting this population's needs for environmental and climate change education.

A 2016 study by de la Garza on the effectiveness of pedagogical mentorship programs in rural and Indigenous schools found that in these schools,

> multiple cultural and structural impediments, such as inadequate teacher training on [bilingual intercultural education]... [and] viewing Indigenous languages as a problem... prevented linguistically and culturally relevant education from being enforced... (de la Garza 2016, p. 54)

The lack of relevant bilingual educational offerings is potentially the biggest area of concern when it comes to expanding access to climate change content for Indigenous populations, as even teachers who attend Bilingual Normal Schools for their pre-service teacher education report feeling mostly unprepared to teach the subjects within the Basic National Curriculum (de la Garza 2016). Since many Indigenous communities are located in parts of the country that are immensely susceptible to climate risks and this population is likely to be among the most vulnerable to the impacts of climate change, it is essential to consider the issue of inadequate bilingual education when considering interventions to creating resilient, climate action-oriented communities.

3.4.2 *Out-of-School Youth*

A second gap identified in reviewing Guatemala's existing climate change education efforts is the country's large, out-of-school youth population and the limited access this population has to information on climate change. In order to understand this challenge in context, it is important to note that the Guatemalan constitution, enacted in 1985, guarantees freedom of education and establishes the obligation of the State "to provide and facilitate education to its inhabitants" (Congreso de la República, 1985, p. 25). More specifically, Guatemalans have "the right and the obligation to receive [free] initial, pre-primary, primary, and basic education within the age limits established by the law" (Congreso de la República 1985, p. 26).

In the more than three decades since the establishment of free basic education in Guatemala, the average length of schooling for youth in the country is still extremely low, at 6.5 years (UNDP 2018). According to USAID (2018), there are currently 1.7 million out-of-school youth, aged 15–24, in the country, many of whom failed to complete a primary education and have limited options for entering the workforce. More than 660,000 of these young people are located in the country's Western Highlands, a majority-Indigenous region where 60% of the population lives in poverty and which is among the world's most vulnerable areas to climate impacts (Rainforest Alliance 2017).

Within the aforementioned National Policy on Environmental Education, the government acknowledged that this is yet another societal issue the country has yet to address in its efforts to promote environmental education (MARN 2017). Despite this acknowledgment, there is very little evidence to suggest that it has focused any effort on targeting this population for provision of CCE. Instead, the evidence suggests that the government's efforts have largely been focused within the country's formal education system and specifically on reforming the national curriculum to include these topics. Based on this lack of information regarding efforts targeting out-of-school youth, and based on the fact that a large portion of this population is largely Indigenous and lives in some of the world's most vulnerable regions, we believe that there is yet more reason to be concerned about the equitable distribution of CCE in Guatemala.

3.4.3 Lack of Coherence and Alignment Between Different Components of the School System and Climate Change Education

A last major gap identified in Guatemala's CCE efforts is the lack of interdisciplinarity in curriculum content on the topic, as well as a general lack of cohesion between this curriculum and other components of school systems. As previously discussed, the Guatemalan Ministry of Education has made the incorporation of climate themes into the country's Basic National Curriculum a priority. However, a detailed review of this curriculum reveals that climate change-related competencies are only targeted within its natural sciences section, despite research on education for sustainable development that stresses the need for an interdisciplinary approach to these topics (Ministerio de Educación 2007; Annan-Diab and Molinari 2017). Multifaceted issues such as climate change, Annan-Diab and Molinari argue, "require knowledge and skills from distinct disciplines in an integrated manner. Interdisciplinarity promotes the ability to understand complex problems and act on them." (Annan-Diab and Molinari 2017, p. 11).

Emphasis on climate change topics and the need for this type of interdisciplinary approach is also largely missing from other aspects of the education system, including its teacher professional development structure. Despite advocating for a vision of education that "promote[s] and develop[s] the ideals for a sustainable Earth: a world that is just, with equity and peace, in which individuals take care of the environment to contribute to achieving intergenerational equity," (Universidad de San Carlos Guatemala 2009, p. 5) the national in-service teacher training program, run through the University of San Carlos, has largely failed to reflect this vision. The training program reflects the same lack of interdisciplinarity towards environmental topics that is present in the Basic National Curriculum, with these topics only being mentioned within the natural sciences portion of the teacher preparation plan. Further, the plan includes no mention of preparation for teachers in topics specifically related to climate change, global warming, or related subjects (Universidad de San Carlos Guatemala 2009).

Despite these concerns regarding the content of the teacher professional development program, an even bigger concern may be the number of teachers the program is actually reaching. A 2011 study found that, when given the same assessment as their students, teachers in sample schools in Guatemala were unable to answer 80% of questions related to the topics they were expected to teach their students (Rojas 2011). In a separate survey, only 18% of teachers in the Western Highlands of the country reported having received in-service training on curriculum topics (Rubio et al. 2017). While the Ministry of Education has dedicated special efforts to incorporating climate change into the curriculum, it has failed to not only acknowledge its complex and interdisciplinary nature, but also to reinforce other components of its education system in an effort to ensure that these curriculum efforts can be effective.

3.5 Moving Forward with a Solution

In light of the multipronged nature of the causes of the limited extent of climate change education in Guatemala, we considered several options to address these deficiencies. In order to address the lack of bilingual education, we explored the possibility of creating early literacy Indigenous language instruction materials focused on environmental themes. Similar to Bazin and Saintis in Chap. 4 of this book, we also recognize the great need to educate out-of-school populations in Guatemala and thus also considered creating a leadership development program focused on the theme of leadership for climate action. Finally, in an effort to remedy the lack of cohesion and interdisciplinarity towards climate topics in the school system, we investigated developing a guidebook utilizing a whole-school approach to creating a culture of climate action within schools.

After considering each of these three alternatives and evaluating these different programs ideas using criteria such as efficiency, equity, sustainability, and feasibility, we concluded that working within the formal education system and using a whole-school approach to climate change education was the most promising option to create a culture within communities of shared responsibility towards the environment and action towards mitigating the effects of climate change. While we discovered glaring gaps in the provision of climate change educational topics to certain groups, including Indigenous communities and out-of-school youth, our criteria indicate that the most realistic and feasible problem for us to address at this time is the lack of interdisciplinarity and cohesion between different school elements in the provision of these topics in formal school settings. Additionally, we feel that while programs targeted towards marginalized groups address important symptoms of the problem we have identified, the whole-school approach more directly targets the root cause of the problem, which is the lack of a culture of shared responsibility towards the environment and fellow citizens in Guatemalan society.

Ultimately, our hope is that by presenting schools with a focused resource to guide them in incorporating the idea of climate action into different components of school life, these different components will continually reinforce and strengthen each other in a way that will ultimately create a culture of climate action within the school environment and subsequently, in individual communities and in the country as a whole. As school leadership focuses on sustainable practices and improving the school culture, for instance, teachers are supported to incorporate such lessons into their own teaching, using innovative and impactful practices, all while students are encouraged to learn these topics through project-based learning and by engaging in relevant climate issues with local organizations. As each of these components work together towards the goal of improving the local environment, students will become engaged and motivated to make change outside of school, and a culture of climate action will emerge.

3.6 Preparing a Whole-School-Centered Guidebook for Schools

Given that the goal of providing schools with a resource to guide them in implementing a whole-school approach will be the most effective in creating a culture of shared responsibility and climate action in Guatemalan communities, how can existing resources and best practices on this kind of approach be incorporated and implemented within a school? In the next section we present a review of UNESCO's *Getting Climate Ready* (Gibb 2016) report in an effort to identify its most relevant strategies for the Guatemalan communities and schools our guidebook will target. Concurrently, the section will review additional literature on the global "green school" movement in an attempt to identify and consider additional applicable elements to the whole-school approach not discussed within UNESCO's report. As these elements are identified, each is then elaborated on using an understanding of current best practices and educational research on its effectiveness. Finally, we present a set of proposed activities for a pilot school in Guatemala to implement, alongside expected outcomes for these activities and their relevance to the overall purpose of our project.

UNESCO's *Getting Climate Ready* (Gibb 2016) report draws upon extensive research on the effectiveness of adaptable models of school organization in Education for Sustainable Development (ESD), as well as on work being done by schools around the world to take action against climate change. Specifically, it complements this research by suggesting an array of specific actions schools can take to build climate resilience. The report explains that "the whole-school approach to climate change means that an educational institution includes action for reducing climate change in every aspect of school" (Gibb 2016, p. 3). In an effort to create a culture of sustainability within schools, the report presents general guidelines for school governance, teaching and learning, community partnerships, and facilities and operations. These guidelines include ideas such as ensuring everyone in the school has a role to play, addressing climate change in an array of subject areas, providing opportunities for critical, creative, and future thinking, and enabling students to take action (Gibb 2016).

The feasibility and effectiveness of UNESCO's approach can first be assessed theoretically through the general lens of research done into both large-scale, system-wide reforms, as well as into smaller-scale approaches to school organization and improvement. On the one hand, the literature on successful system reforms find that similarly to the whole-school perspective, such reforms combine both alignment and coherence, meaning that learning, "is the goal of the various components of the system… [and] that the components reinforce each other in achieving whatever goals the system has set for them" (World Bank Group 2018, p. 13). Likewise, school organization and improvement literature ultimately challenges the role of traditional pedagogies in promoting global sustainability and suggests that instead, more adaptable models of school organization that respond at all levels to new research and social movements are better suited to realizing the Education for Sustainable Development (ESD) vision of creating aware and responsible individuals (Morgen et al. 2019). Notably, it has been found that schools which adopt an interdisciplinary, multidimensional ESD implementation strategy have more supportive organizational structures, are better able to leverage community relations, and are ultimately more effective in translating ESD's objectives into practical reality (Morgen et al. 2019).

Complementary to these findings are the more practical results that have emerged from schools around the world operating under the umbrella of what some call the "green school," or "eco school," movement. These green schools have adopted the kind of whole-school approach advocated for in UNESCO's guidebook, and "aim to include everyone (students, teachers and the local community)… to motivate students to take on environmental problems and seek resolutions particularly at a local level but also thinking globally…" (Gough et al. 2020b, in press). To that end, schools within these networks focus on developing environmental policies and strategies, providing teacher professional development, involving key stakeholders in sustainability decision-making, and other related actions (Gough et al. 2020a). While measuring the broader impacts of these schools has been difficult, many countries have reported evidence of the development of more sustainable practices, an increase in ESD curricular content, increased student action in influencing government policy, and more (Gough et al. 2020a).

Unfortunately however, little additional information exists on the long-term impact of green schools centered primarily around climate action, especially in developing countries contextually similar to Guatemala. Regardless, short-term evaluations of green schools completed in other developing contexts do provide some hope that a similar approach might be able to take root and have an impact in the country. For instance, a recent evaluation of an Eco-School network in the Small Island Developing Countries (SIDS) of Mauritius, Madagascar, Comoros, and Zanzibar found that such an approach was able to foster positive attitudes towards the schools' role in equipping individuals to address critical challenges by inspiring changes in teaching methods and emphasizing place-based and problem-based learning. Additionally, evaluators observed physical improvements in the schools' grounds and in their wider communities, as well as improvements to key student outcomes such as school attendance (Copsey 2020).

In light of this additional research on school organization and the green school movement, we have concluded that while the recommendations included in UNESCO's report constitute a sound basis for thinking about preparing schools to meet the climate challenges facing their communities, there is a critical need to enhance these guidelines to suit the specific needs and opportunities of Guatemalan schools. Furthermore, in an effort to make this guidebook even more context-specific, our enhanced recommendations, presented in the following sections, will be tailored to a school we have identified in Guatemala's Western Highlands, Atitlán Multicultural Academy (AMA).

AMA is located in a largely Indigenous community located within the department of Sololá, which in the last few years has suffered the impacts of climate change primarily in the form of excessive rainfall and severe droughts (Asociación Amigos del Lago de Atitlán 2018). The school itself is a "K-12th grade, English-immersion school serving students from all backgrounds representative of the Guatemalan Highlands" (Atitlán Multicultural Academy 2020a), with a population of no more than 100 students and a curriculum that strongly emphasizes project-based learning (Atitlán Multicultural Academy 2020b). Our reasons for selecting AMA as our target school include the region's elevated vulnerability to climate change, as well as the added feasibility afforded to us by the school's instructional language, existing curriculum and partnerships, and small student population.

As an additional note before presenting our final recommendations; we want to emphasize the importance of two-way communication in crafting this final product. As authors of these recommendations, we have communicated our thoughts and rationale with AMA and have received valuable feedback from faculty and staff in response, much of which has been incorporated into our suggestions. The success of a program like this can only be possible if the school can communicate what works for them and is encouraged to make adjustments as necessary. Schools work hard to address a multitude of priorities, so it is our hope that the whole-school approach can be integrated into systems and structures that schools already have in place.

Additionally, we want to make clear that this is just the beginning of an iterative process. Again, for a program such as this to be successful, it is crucial that our partnership with the school remain open and ongoing in order to make continued necessary adjustments so that implementation of the following components can be achieved, and to ensure continual support so that the whole-school approach can be sustained. As the school begins to implement various aspects of this model, we acknowledge that further revisions will be necessary and we remain committed to eliciting feedback from the school throughout the process in order to create the optimal approach. With that in mind, the following sections present an early-stage version of the recommendations for establishing a whole-school approach based on feedback from members of the AMA community along with some example accommodations that other schools may also consider in implementing the whole-school approach to climate change education.

3.7 School Leadership

The first major element of a whole-school approach addressed within the UNESCO guidelines is the role of school leadership in creating a school culture of climate action. As Gibb (2016) explains, the kind of culture this approach aims to cultivate is one in which responsibility for achieving a schools' climate action goals are distributed at all levels and not simply concentrated at the top. This form of equitable, distributed leadership aligns with the argument that the successful promotion of sustainability through education necessitates the involvement and action of people in different positions who embrace the idea of change (Wals and Benavot 2017). Specifically, the idea has been shown to effectively engender a school culture that is conducive to conversations on climate action and sustainability, and supportive of related initiatives (Morgen et al. 2019).

In practice, this type of governance structure has already been implemented in green schools around the world and is actually one of the most critical components of the Foundation for Environmental Education's (FEE) Eco-Schools framework. Like in UNESCO's guidelines, the creation of an Eco Committee is the first step in FEE's program, and aims to be as representative of the whole school as possible. According to FEE, the purpose of these committees is to ensure that the voices of all members of the school community are heard, and to ensure that the other facets of the program are successfully implemented (*How FEE EcoCampus Works* n.d.). Reviews of FEE's specific approach have found that such committees help to address one common concern related to the implementation of school-wide sustainability programs, which is the fact that they are typically introduced by a science teacher and thus become disintegrated from other disciplines in the school (Lysgaard et al. 2015). Involving different disciplines and actors of the school community in a student-led committee and giving them ownership over its initiatives allows schools to tackle this problem (Gough et al. 2020b).

In light of this research, our guidebook incorporates the distributed leadership strategy and largely resembles FEE's and UNESCO's frameworks for implementing this key aspect of the whole-school approach. Within this component, the major activity we advocate for is the creation of a Climate Action Committee to coordinate the development and implementation of the school's climate action plan (Gibb 2016). More specifically, the role of this committee would be to plan for the school's climate-related projects, collaborate on their execution, and work throughout the year to assess their progress and adjust activities accordingly (Gibb 2016). In terms of actually creating the committee, school leaders must ensure to recruit individuals from both inside and outside the school who might have a stake in its climate activities, including students, teachers, administrators, custodial staff, and local community leaders. Finally, leaders must ensure that the group of individuals participating in this activity are also representative of the diversity in age, gender, race, and socioeconomic background of the greater school population (Gibb 2016). Figure 3.1 shows a guidebook recommendation for the first stage of this activity, addressed specifically to school leaders.

Creating a School Climate Action Committee:

In terms of the actual creation of the committee, you must ensure to recruit individuals from both outside and inside the school who might have a stake in its climate activities (Gibb 2016). For this reason, the first step you should consider taking should be the drafting of what is called a stakeholder map. In simple terms, this entails identifying the different groups of individuals within the school that might have a "stake" in its climate action activities, and determining their specific interests in these activities, as well as their level of influence within the school and community. Creating this map will secure representation from different stakeholders in the committee, ensure that the committee will have access to different skills, knowledge, and perspective, and provide insight into potential roles within the committee (Gibb 2016).

Important stakeholders are students, whose roles could include planning and leading projects or mentoring younger students on how to take part in climate action. Teachers, administrators, cafeteria and custodial staff, and local community members and organizations are other examples of important stakeholders and potential Climate Action Committee members. One final, important piece of the recruitment aspect of this activity is ensuring that the group of individuals participating in this activity are also representative of the diversity in age, gender, race, and socio-economic background of the greater school population.

Fig. 3.1 How to create a School Action Committee as proposed in UNESCO's guide to whole school climate change education

In speaking with AMA staff about this proposed activity, it was raised to us that the biggest challenge to its successful implementation at the school would be the time commitment required from all parties involved, particularly from the school director. This notion is particularly concerning given existing research on challenges to the implementation of a whole-school approach in green schools, which finds that "programs can often not succeed or be sustained when there is a lack of ownership of the program by the whole school community, or when the program leaders in the school burn out or leave" (Gough 2020, in press). Therefore, with an understanding that success of the whole school approach and of this activity in achieving their intended outcomes relies primarily on the assumptions that there will be active and sustained interest from all stakeholder groups, and that individuals will have the ability to commit the amount of time necessary to carry it out, we found it important to consider ways to then align our recommendation with structures and activities already in place within this specific school.

One potential accommodation that resulted from these conversations is for AMA to incorporate the idea of a Climate Action Committee into its already existing club program. Within the school, students currently meet for one hour per week, during school hours, in interest-based clubs with other students from different grade levels and a classroom teacher in order to undertake activities related to this interest. Adding the Climate Action Committee as a club option for students to join at AMA will free the school director and teachers from the burden of having to schedule time outside of school hours for the committee to meet. Students, too, will be more likely to participate in the committee and be engaged if it is an activity they were able to opt into and one that does not require them to stay past regular school hours.

Once this club or committee is created, our expected output is the presence of an active Climate Action group in the school, within which leadership and responsibility for the school's climate goals are distributed in a representative manner, and within which concrete action is being taken towards achieving these goals. The main indicator we propose for determining that this output has been successfully achieved is that the committee has decided what these climate action goals are and within a year of its creation has carried out a predetermined number of activities in the pursuit of these goals. The means of verification for this indicator will come from reports drafted by a designated member of the committee detailing the committee's activities for the year, including attendance numbers and cost summaries.

If the Climate Action Committee is created as planned and successfully executes its activities, and assuming that attendants to these activities see the value of participating and feel motivated to replicate these actions in their wider community, then this committee will serve towards our larger purpose of creating a community of individuals who understand their responsibility towards the environment and use their skills and knowledge to take action against climate change. Additionally, its successful implementation will be vital in reinforcing the other components of the whole school system. Involving teachers and students in this committee, for instance, should make them feel more invested in participating in the professional development and curriculum aspects of our program.

3.8 Community Partnerships

Another aspect UNESCO identifies as crucial to the whole-school approach is the formation of partnerships between schools and community organizations. Through such partnerships, students are able to both learn about the climate and apply said learning outside of the classroom. These relevant, hands-on learning experiences can help students feel better connected to their communities and can create an experience that is a "more effective and long-lasting form of learning" (Beard and Wilson 2006, p. 1). Research indicates that as a result of these experiences, students may be more likely to take action towards solving real problems their communities are facing (Karpudewan and Khan 2017).

Community partnerships also present potential opportunities for professional development for teachers. For example, Marlow and McLain (2011) found that teachers reported a multitude of benefits when participating in experiential learning experiences within various community settings themselves. Experiential learning for educators can, much like for students, serve as "potentially transformative experiences and provide numerous opportunities to touch teachers in unique and highly personal ways" (Marlow and McLain 2011, p. 9). Such experiences for teachers also help to address any worries about implementing these methods in their classrooms (Girvan et al. 2016). In this case, experiential learning in conjunction with local organizations could potentially increase teacher comfort with incorporating climate-action objectives into their curriculum. As a result, schools observe increased implementation in classrooms (Girvan et al. 2016).

Upon review of whole-school approaches to sustainability worldwide, Henderson and Tilbury reported that "partnerships are key components of program design and implementation and in many cases are seen as critical to the program's success" (2004, p. 19). These partnerships vary, in that some partnerships exist between a school and government authorities, while other schools may partner with civil sector organizations. Additionally, schools' motivations for establishing partnerships may differ. Some partnerships may be intended to increase financial support, while other partnerships may serve as vehicles to serve the local community (Henderson and Tilbury 2004). New Zealand's *Enviroschools* Program utilizes partnerships with a variety of stakeholders. Through their partnership with the Ministry of Education, for example, *Enviroschools* are able to "strengthen the capacities of teachers and professionals to work effectively in [environmental education]" (Henderson and Tilbury 2004, p. 21). Partnerships create valuable opportunities for schools committed to mitigating climate change.

A partnership already exists between AMA and local organization, Amigos del Lago. The organization is a non profit association that works to "educate, research and ensure [Lake Atitlán's] conservation" (Asociación Amigos del Lago de Atitlán 2018). Amigos del Lago has already helped to enhance environmental science lessons at the school through providing hands-on activities, and representatives from the organization have previously assisted with the school science fair as judges. We see that this partnership has the opportunity to expand, and is also an indication of the school's willingness to partner with other organizations.

With this in mind, our guidebook recommends that schools should partner with various local organizations that share "a common vision" with the school and that prioritize sustainable practices, (Blank et al. 2012). Through these sustained partnerships, an array of hands-on and authentic learning experiences can be offered to students in a manner that "encourage[s] each stakeholder to clearly define its role in meeting specific goals" (Blank et al. 2012, p. 13). Rather than simply suggesting schools "have students learn through experience," (Gibb 2016, p. 16) we provide information about specific local organizations who provide hands-on learning experiences in and out of the classroom for students throughout Guatemala.

Given AMA's preexisting relationship with community organizations, recommendations regarding this portion of the whole-school approach was well-received by the staff. In fact, faculty members were in the process of initiating additional partnerships, specifically with organizations that could help make clear the importance of advocacy when it comes to matters of climate change in the region. For schools that may not be located near community organizations such as Amigos del Lago, we suggest they look to local colleges and universities, who may have courses or student-led organizations with climate-oriented missions, just as Lee and Nam suggest in Chap. 6. Yet another option may be to reach out to organizations whose missions may not be directly tied to the environment, but who value sustainable practices, such as markets that use sustainable practices.

For those schools just beginning the process to establish partnerships, a reality exists that challenges may arise during these efforts. For example, as experienced by the Eco-Schools Programme in Kenya, "it takes a long time to build relationships with partners due to different priorities and bureaucracy, thus slowing down effective project implementation" (Otieno et al. 2020, in press). This can be particularly problematic for schools who rely on these partnerships to provide resources associated with other components of the whole-school approach, such as professional development or curriculum materials. Other schools have found the "lack of support for community partnerships and competition between various initiatives within the school" (Gough 2020, in press) to be additional challenges in this realm.

Once schools have initiated a partnership with a local organization, and our assumptions are met, we expect the outcomes of the sustained partnership to include increased student engagement within their communities and improved skills among students in regards to climate action. The partnership serves the organization as well, as working with schools generates increased support for the organization's work and contributes to their work in combating climate change. Indicators by which schools can evaluate their progress within this aspect of the whole-school approach is the number of community partnerships initiated. Schools should also hold themselves accountable to maintain the partnership through engaging their students with the organization, which can be verified through written agreements between the school and organization. Ideally, within the first year of establishing this whole-school approach to becoming a climate action-oriented school, schools should be able to report at least one community partnership.

In an effort to provide a guidebook that is relevant in the sense that it is context-specific, the community partnerships section profiles local organizations for local schools to reach out to. The organization's name and information is provided, along with a brief overview of some of the organization's work and any school programs they have in place. Figure 3.2 provides an example of what a community partnership profile might look like in the guidebook.

If community partnerships are created and sustained in a way that allows students to interact and become engaged with the community, assuming students see the value of the work they are doing, then these community partnerships will contribute to the overall purpose of creating a community that uses their skills and knowledge to combat climate change and understands their shared responsibility to protecting the environment. As previously stated, these partnerships will also aid in advancing the other components of the whole school approach. Community partnerships provide unique opportunities for quality teacher development and infuses environmental lessons into the school's curriculum in multiple ways.

Amigos del Lago

Website: https://amigosatitlan.org/home/

The Work: Amigos del Lago works to educate the community about the lake and the importance of conserving it for cultural and environmental reasons.

School Programs: This organization is familiar with working with schoolchildren, as they work in over 200 schools in the department. Their programs are designed to specifically address the local context. School programs include exercises that focus on sustainable practices and habits. A reference manual with resources and activity ideas specific to the region can be downloaded on their website.

Fig. 3.2 An example of a community partnership to support whole school climate change education

3.9 Curriculum

Curriculum is another important factor in the creation of a whole-school approach to combating climate change. Namely, UNESCO (2017) suggests that schools identify core competencies their students need in order to work towards sustainable development. Schools are then encouraged to incorporate these competencies into the curriculum as it already exists. These objectives are divided into three domains: cognitive, socio-emotional, and behavioral (UNESCO 2017). Respectively, some competencies outlined in these objectives include: knowledge of how human actions contribute to the promotion or prevention of climate change, collaboration towards fighting climate change, and the evaluation of the effects of actions on the climate (UNESCO 2017).

Oldakowski and Johnson (2018) determined that an integrated curriculum, specifically a curriculum that incorporates matters of climate change into core subjects such as math, "leads to improved learning outcomes in the short-term for all…subjects." (p. 22), and concluded that content-specific skills supported climate-action skills, and vice versa. Results also indicated that schools with great demographic, socioeconomic, and academic diversity among their student populations "improved at the same and sometimes greater magnitude" than their counterpart schools, when given the opportunity to learn in an integrated setting (Oldakowski and Johnson 2018, p. 22).

This integrated approach to curriculum can be seen modeled in green schools around the world. In Israel, one component the Ministry of Environmental Protection (MEP) and Ministry of Education (MoE) look for when certifying a school as a Green School is an integrated curriculum (Tal 2020). In this way, Green Schools in Israel ensure that "environmental topics and concepts are included in various subjects" (Tal 2020, in press). This approach to curriculum can also be seen in Green Schools located in India. Integrating environmental education into other subject areas aids in "ensuring that adequate time is earmarked for pertinent activities" (Sharma and Kanaujia 2020, in press), while also presenting the content in a relevant way.

Our target school, AMA, acknowledges the importance of thinking in an interdisciplinary manner. Students are encouraged to ask questions in ways that can only be answered by drawing upon knowledge from multiple subject areas. The school also conveys to students that knowledge that spans across subject areas empowers and enables them to make positive change in the world (Atitlán Multicultural Academy 2020a). AMA prioritizes creating opportunities that enable students to construct their own knowledge through the use of a hands-on, project-based curriculum. This approach to teaching and learning, paired with the volunteer opportunities provided to students, creates AMA's "Whole People, Whole Planet" approach, which aims to foster a "wide and inclusive world view" in students (Atitlán Multicultural Academy 2020a).

Given the research and the way in which AMA values interdisciplinarity, this section of our guidebook provides sample lesson plans which model the integration of climate change matters into content area lessons. The previously mentioned objectives for climate change were used as a guide, along with climate matters of great relevance to Guatemalan communities, in order to unite lessons in climate-action with content-specific work. We additionally encouraged teachers to provide students opportunities to develop and improve the cognitive competencies necessary for sustainable development through the study of local environmental concerns. In this way, teachers may address multiple content area standards, along with major objectives addressing climate action.

In speaking with teachers at AMA, we found that many classrooms were already working to incorporate themes of climate change into content area lessons. However, like many teachers in an array of contexts, the biggest roadblock to carrying out such integrated lessons was the lack of appropriate resources. AMA teachers aim to deliver engaging, relevant, and developmentally appropriate lessons to their students. However, finding resources specific to the region's climate that are appropriate for the age group is challenging. Therefore, it is very important for the guidebook to include not only sample lessons, but also accompanying materials, such as those presented by Rhodes and Wang in Chap. 2. In fact, we suggest schools working to introduce the whole-school model look to Chap. 2, as well as Japanwala's work in Chap. 5 for further assistance in the realm of curriculum development.

Gough's review of the challenges green schools face highlights this challenge. Existing curriculum materials that schools may have access to "often do not support green school programs" (Gough 2020, in press). Schools embarking on the journey to adopting a whole school approach to sustainability should be cautioned that it is often the lack of such resources that lead to burn out (Gough 2020). While this guidebook encourages teachers to integrate climate themes into their lessons as they see fit, it is crucial that schools do not burden their teachers by doing so, leading to "eco-fatigue" (Gough 2020, in press). As schools work towards creating an integrated curriculum as part of their whole-school approach, they should be acutely aware of these challenges.

The curriculum section of the guidebook provides several sample lesson plans. These plans are intended to model the creation of integrated lessons, incorporating subject matter content with climate-action objectives using relevant climate topics.

Figure 3.3 provides just a portion of one example lesson plan from the guidebook. Not included in this example is the detailed sequence of activities, which provides teachers step-by-step plans for carrying out such an integrated lesson.

In order to achieve this integration of climate-action skills into the school's curriculum, several assumptions must be met. Specifically, in order for this integrated curriculum to be effectively carried out in schools, we assume that teachers will be willing to address climate themes in their lessons. There is also an assumption that teachers will feel motivation towards creating and carrying out climate-related lessons. The successful integration of this curriculum also relies on an assumption that students will willingly participate in these types of lessons.

Once climate-action competencies are integrated into the school curriculum, and our assumptions are met, our expected outputs include improved student content knowledge about the climate, increased knowledge about sustainable practices, and a rise in student engagement. We propose that the success of the outcomes be measured by student growth in the area of climate change content knowledge. As a means of verification, we suggest schools use data from pre- and post-tests in order to track student improvement in knowledge about climate change and sustainability.

If schools and teachers have access to climate change related activities and lesson plans, and teachers make use of these resources effectively in their classrooms, assuming students are positively impacted by and engaged with the material, this

Grade 4 Math - Deforestation & Reforestation

Overview:

Forest pests, diseases, and forest fires have been affecting the forest areas in Guatemala for decades. Long periods of drought have increased the possibility of forest fires in recent years as well. Yet the forest provides many positive benefits for the Western Highlands. Trees serve as shelter for animals, provide jobs for people, and absorb pollution and produce oxygen. For these reasons, and more, it is important that we minimize deforestation and plan for the planting of trees to replace forest areas previously lost.

Student Learning Objectives:

1) Students will be able to determine the area of a given space (in the shape of a quadrilateral) that has experienced deforestation.

2) Students will be able to apply their knowledge of area to reforestation efforts, in order to determine the number of trees that need to be planted in certain areas.

3) Students will be able to advocate for change to improve their local environment.

Assessment:

Observe students during work time to evaluate accurate use of strategies for computing area. Use student calculations and presentations for continued evaluation of proper understanding of area, as well as an ability to advocate for positive climate change.

Fig. 3.3 An example of a lesson plan in whole school climate change curriculum in Guatemala

integrated approach to curriculum will contribute to the overall purpose of the program of increased understanding of responsibility for the environment and use of knowledge and skills to fight the effects of climate change. Moreover, an integrated curriculum will reinforce the other aspects of the whole school system in a way that makes the message of shared responsibility clear. An integrated curriculum, in particular, will support student work with community partners in a meaningful way.

3.10 Teacher Professional Development

One final dimension mentioned within the UNESCO guidelines is that of teaching and learning. The guidelines acknowledge that addressing climate change is complex and involves addressing not simply scientific and environmental issues, but also social, cultural, and political issues (Gibb 2016). For this reason, and as previously discussed in the section on curriculum reform, schools should aim to integrate climate action into all subjects and help students develop critical thinking skills to enable them to address these issues (Gibb 2016). Yet, simply infusing new content into the curriculum will be meaningless without one additional, critical action not discussed in the UNESCO guide: adequately preparing teachers to teach these topics and skills in dynamic ways (Gough 2016; Karrow and DiGiuseppe 2019).

While not a component of UNESCO's approach, it has generally been shown that the promotion of good teaching methods clearly has a significant impact on student outcomes, and that these methods develop as a result of the acquisition by teachers of specific content knowledge, the promotion of certain ethical values and attitudes, and the development of practical skills under the guidance and supervision of experts (Villegas-Reimers 2003). In regards to environmental education specifically, the importance of reaching teachers through pre-service and in-service environmental education programs has long been understood, with UNESCO itself declaring as early as 1977 the need for teachers "to understand the importance of environmental emphasis in their teaching" (UNESCO 1978, pp. 35–36).

For this reason, schools committed to effectively instituting a whole-school approach on climate action must therefore make an intentional effort to equip teachers with the skills and knowledge necessary to effectively carry out their responsibilities within this model. In fact, many environmentally-minded schools around the world are already stressing the professional development of teachers as an important component of ESD, in order to develop and improve educator competencies around teaching climate change (Henderson and Tilbury 2004; Gough 2016). The Eco-Schools program in Malawi, Uganda, and Tanzania, for instance, leverages partnerships with local education authorities and organizations to deliver intensive training for teachers on the Eco-Schools concept and methodology. A recent evaluation of this program completed by the Danish Outdoor Council found that these activities resulted in an increase of both teachers and students applying new environmental learnings at school and at home, an improvement in teaching techniques, and improved academic performance (Danish Outdoor Council 2017).

Within the context of AMA, teacher professional development takes several different forms. Before the beginning of each school year at AMA, new and returning teachers participate in a two-week teacher preparation period, during which they each create and present their own professional development workshops to the rest of the faculty, receive guest lectures from community leaders on topics pertaining to pedagogy and curriculum content, and are given time to collaborate on curriculum development and lesson planning (personal communication, December 7, 2019). During the school year, teachers are then paired up and encouraged to visit each other's classrooms and give feedback. While there is certainly room for improvement in terms of continued professional development for teachers throughout the school year, the school has shown a commitment to supporting the improvement of its teachers in a variety of ways (personal communication, December 7, 2019).

Based on both the research supporting teacher professional development in CCE and the supportive culture built around teacher improvement at AMA, we intend for our guidebook to stress the importance of teacher professional development as a component of the whole-school approach. In our attempt to do so, we propose several activities for the school to implement in order to develop all teacher's competencies in climate change topics, not just science teachers. Our first recommendation is for schools to have teachers regularly participate in in-service training workshops on environmental and climate change education. The topics of these workshops can range from understanding the science behind climate change, to guidance on incorporating climate change topics into non-science subject lesson plans.

In speaking to the staff at AMA, we once again received the feedback that the biggest challenge to the implementation of these workshops will be the time requirement for teachers. Currently, staff meetings are held only once per week for 45 minutes at the school, meaning that any workshops that take place during this time would have to be quite short. The staff agreed that teachers would be reluctant to attend additional workshops outside of this time. For this reason, our guidebook still provides workshop recommendations, but it also stresses additional teacher PD activities that will have the aim of allowing teachers to become active participants in their own development and that of their colleagues without adding to the hours they need to work. An example of how one such recommendation will look is in Fig. 3.4.

As with the previous elements we have discussed, the success of these activities in achieving our preferred outcomes relies on several assumptions we are making about the teachers that will be participating in this professional development. The first big assumption we are making is that teachers will see the value in additional professional development opportunities specifically focused on climate change topics and feel invested in increasing their knowledge on these topics. An additional assumption is that both teachers and the administration will be on board with changes in pedagogy that might result from these opportunities, namely, a shift to more project-based learning activities in the classroom. Finally, one last, large assumption is that once teachers have learned new skills and knowledge related to teaching climate-related topics, they will actually feel comfortable in delivering this content in the classroom.

Creating a Peer Feedback System Within Your School

A potential complement to climate change teacher professional development workshops is the creation of a peer feedback system, which will allow teachers to invite one another into the classroom for informal observations and create the opportunities for feedback on the incorporation of climate topics into different subject area lesson plans (Gonzalez, 2016).

One easy and convenient way to set up this system is through the creation of a Pineapple Chart. This is a chart that can be posted somewhere within the schools where teachers go on a daily basis, such as the teacher's lounge or supply office. On the chart, which may look like the weekly school schedule, teachers will be able to fill in time slots with the climate change-related activities they are planning to carry out in their classes at these times. Other teachers who stop by to see the chart are then able to "sign up" to go and observe said activity and then provide feedback on how they think it went (Gonzalez, 2016).

This system can be as formal or informal as the teachers in your school decide they want it to be. In an informal system, teachers can decide to pop into the activity for five minutes, or for the entire hour. The provision of feedback may come in the form of a formal write-up, or in a quick chat during a planning period (Gonzalez, 2016). The point of the activity is to create a system that teachers are comfortable with, and one that will make them feel excited about enhancing their integration of climate topics into their lesson plans through giving and receiving help from their peers.

Fig. 3.4 A professional development system to support whole school climate change

Once teacher professional development activities are implemented in schools and our assumptions about teacher motivation and comfort are met, our expected outputs include increased teacher content knowledge of climate change and increased incorporation and delivery by teachers of climate change-related topics in lessons in all subject areas. The main indicator we propose for determining whether these outputs have been achieved is that teachers are indeed increasing their delivery of climate-related lessons and are doing so in a competent manner. The primary means of verification for these indicators would be classroom observations and teacher evaluations by the school principal, who would first verify that these lessons are being delivered, and then determine whether it is being done skillfully.

As with the previously discussed components of the whole-school approach, teacher professional development is vital to the eventual achievement of our project's larger purpose. If the teachers' content knowledge on climate change increases and they incorporate this knowledge into lessons in the classroom, assuming that students feel more engaged with the material as a result of their teachers' increased comfort, then this will also serve towards our larger program purpose. Further, increased teacher PD and consequently, increased incorporation of these topics into classroom lessons will also serve to complement the whole school's other components. Increased teacher content knowledge, for example, will allow teachers to be more engaged and productive members of the distributed leadership structure. Teachers who are members of the Climate Action Committee will be able to make more informed recommendations regarding the committee's activities and be able to better guide and support the rest of the members in their pursuit of established climate goals.

3.11 Conclusion

Climate change education is increasingly becoming one of the world's best hopes for preparing individuals to adapt to and mitigate the effects of climate change impacting their communities. While the Guatemalan government has acknowledged its importance and made significant progress in incorporating climate change topics into its national curriculum, the country has yet to achieve a culture of shared responsibility and action towards the environment, according to the country's Ministers of Education and the Environment. One of the primary reasons for this is that the government's CCE efforts have largely ignored the functioning of schools as systems whose different components constantly interact and reinforce each other. In response to this failing in Guatemala's CCE efforts, we have created a guidebook, meant to enable schools to create and fulfill their own climate action goals through attention to four key areas of school life: school leadership, community partnerships, curriculum, and teacher professional development. Our ultimate hope is that by incorporating climate action into these different components, they will reinforce and strengthen each other in a way that will ultimately create a culture of climate action within the school environment and subsequently, in individual communities, and in the country as a whole.

Appendix

According to UNESCO (Gibb 2016), the key to a whole school approach to climate action is a commitment to continual improvement. In many ways, this commitment has been reflected in the process of the creation of our guidebook, which has been the result of deep collaboration and continual back-and-forth feedback between the parties involved in its creation. Going forward, this commitment must also be reflected in the implementation of the whole-school approach. This means that schools must be sure to engage in a process of carefully and intentionally planning for, executing, and reflecting on the incorporation of climate action into the four key elements of the whole school (Gibb 2016). Below is a checklist, adapted from one within UNESCO's guidebook, to help schools ensure they are properly engaging in this process.

Climate Action Checklist – Steps to build a culture of climate action in your school

PLAN
School Leadership
☐ Stakeholder Mapping by School Leadership
The first key step in the creation of a distributed leadership structure is a thorough mapping of all actors, both internal and external, who may have a stake in the school's climate action activities. These individuals may include students, teachers, principals, school staff, community members, and others.
☐ Creation of a Climate Action Committee or Club
In the creation of the school Climate Action Committee, school leaders must ensure to recruit individuals from both inside and outside the school who might have a stake in its climate activities, and must ensure that the group of individuals participating in this activity are representative of the diversity in age, gender, race, and socio-economic background of the greater school population.
☐ Assessment of school needs in relation to climate action
Once created, the school Climate Action Committee or Club must collaborate on an honest assessment of the school's current needs and performance in the area of climate action. This will be critical for setting ambitious, yet realistic and feasible priorities for climate action.
☐ Development of a climate action plan for the year
The school's climate action plan should outline the school's priorities and planned activities for the years, state the budget and expected costs for these activities, and lay out their expected outcomes and timelines.
Community Partnerships
☐ Research into local organizations
The Climate Action Committee will research local organizations who have missions to combat climate change. The committee might look into environmental organizations, small businesses, large corporations, or other organizations nearby.
Integrated Curriculum
☐ Assessment of existing curriculum resources and needs
School leadership and the Climate Action Committee will create and administer surveys to all teachers to gather information regarding the existing climate-related resources in the school, as well as the needed resources for incorporating climate-themes into the curriculum.
☐ Conversation with implementing partner regarding outside research and resources
After identifying the resources needed, the committee will share these findings with partners and other stakeholders.
Teacher Professional Development
☐ Assessment of teacher needs, time availability, and existing PD resources
School leadership will assess teacher competencies in climate topics, and administer surveys to all teachers to gather information regarding their needs, interests, and time availability for professional development on the topic.
☐ Conversation with implementing partner regarding outside research and resources
In addition to sourcing recommendations and ideas regarding curriculum integration for local organizations, school leadership should also engage in an assessment of partner capacity to provide resources and assist in teacher PD.

(continued)

ACT
School Leadership
☐ Execution of planned activities for the year
Throughout the course of the school year, the school Climate Action Committee will be responsible for the execution of all planned activities. As part of this execution, committee members will be responsible for collecting information on attendance and costs, along with any other pertinent pieces of information.
Community Partnerships
☐ Establishment of partnerships
Upon identifying local organizations committed to climate action, the committee will reach out to the organizations and establish partnerships. This agreement will be documented so as to keep both parties (the school and the organization) accountable for carrying out agreed upon activities.
☐ Collaboration to plan activities
The committee and the community organizations will plan activities for the school year that align to the needs and goals of both the school and the organization. The activities may include school assemblies, class field trips, or workshops for teachers.
Integrated Curriculum
☐ Teacher integration of climate change topics in their lessons/classrooms
After the appropriate resources have been compiled, teachers will carry out lessons that incorporate relevant matters of climate in their classrooms.
☐ Collection of data on student learning
Teachers will assess student learning and track student growth with regards to climate-related objectives.
Teacher Professional Development
☐ Execution of teacher Professional Development workshops on climate topics
Throughout the year, and as much as is realistic with existing resources and partnerships, teachers should regularly partake in in-service training workshops on environmental and climate change education. The topics of these workshops can range from understanding the science behind climate change, to guidance on incorporating climate change topics into non-science subject lesson plans.
☐ Creation of peer feedback and execution of peer classroom observations/feedback sessions
The creation of a peer feedback system within the school will allow teachers to invite one another into the classroom for informal observations and create the opportunities for feedback on the incorporation of climate topics into different subject area lesson plans.

REFLECT
School Leadership
☐ Gathering and reporting of data collected during each activity
The data gathered by the committee throughout the school year is essential for capturing a change of culture in the school. Such data may include final budgets, attendance at climate action events, and actual school attendance, as well as more qualitative data such as photographs, Climate Action Committee meeting minutes, and more.
☐ Student-led discussion on success of activities and potential for improvement
Have the committee and other school members engage in a student-led discussion around the activities carried out by the committee throughout the year, including a conversation on what worked and what did not.

(continued)

☐ Celebration of success
Plan an end-of-year activity meant to celebrate progress and successes in incorporating climate action into school life.
Community Partnerships
☐ Gathering of feedback on success of partnerships from relevant stakeholders
Create and administer surveys in order to reflect on the success of partnerships. Use this feedback to make necessary adjustments and plan for continued work with partners.
Integrated Curriculum
☐ Gathering of feedback on success of new content integration from relevant stakeholders
Create and administer surveys in order to reflect on the success of integrated lessons. Use this feedback to make necessary adjustments.
☐ Perform evaluations to assess student learning in new curricular topics
Analyze student growth throughout the year with regards to climate change knowledge and skills.
Teacher Professional Development
☐ Gathering of feedback on success of teacher PD from relevant stakeholders
Create and administer surveys in order to reflect on the success of teacher professional development activities. Use this feedback to make necessary adjustments.
☐ Perform evaluations to assess teacher learning in new curricular topics
Analyze teacher development throughout the year with regards to climate change knowledge and skills through teacher reflections and classroom observations.

References

Alcarraz, I.C., Calderón Moreno, L.E., & Barillas, E.M. (2012, January 23). *Documento de Prioridades DIPECHO VII*. Retrieved October 2, 2019, from: https://issuu.com/achcentroamerica/docs/documento_de_referencia__17._documento_pais_guate

Annan-Diab, F., & Molinari, C. (2017). Interdisciplinarity: Practical approach to advancing education for sustainability and for the Sustainable Development Goals. *The International Journal of Management Education, 15*(2, Part B), 73–83. https://doi.org/10.1016/j.ijme.2017.03.006.

Asociación Amigos del Lago de Atitlán. (2018, January). *Educando Para Conservar—Material de referencia para el desarrollo de la educación ambiental en el departamento de Sololá y la Cuenca del Lago de Atitlán*. Retrieved December 13, 2019, from https://amigosatitlan.org/wp-content/uploads/2018/11/educando-para-conservar-2018-br.pdf

Atitlán Multicultural Academy. (2020a). *About Atitlán Multicultural Academy*. Retrieved December 13, 2019, from http://www.ama.edu.gt/about-2/

Atitlán Multicultural Academy. (2020b). *Curriculum*. Retrieved from http://www.ama.edu.gt/curriculum-2/

Beard, C. M., & Wilson, J. P. (2006). *Experiential learning: A best practice handbook for educators and trainers*. London: Kogan Page.

Blank, M. J., Jacobson, R., & Melaville, A. (2012). *Achieving results through community school partnerships: How district and community leaders are building effective, sustainable relationships*. Washington, DC: Center for American Progress. Retrieved from http://www.americanprogress.org/issues/education/report/2012/01/18/10987/achieving-results-through-community-school-partnerships/.

Center for Economic and Social Rights (CESR). (2009). *Guatemala: Rights or privileges? Fiscal commitment to the rights to health, education and food in Guatemala*. Retrieved October 2, 2019, from http://www.cesr.org/guatemala-rights-or-privileges

Centro de Acción Legal - Ambiental y Social de Guatemala (CALAS). (2010). *Decreto Numero 74–96*. Retrieved October 2, 2019, from http://extwprlegs1.fao.org/docs/pdf/gua60658.pdf

CEPAL. (2011, April). *Guatemala: Evaluación De Los Impactos Económicos, Sociales Y Ambientales, Y Estimación De Necesidades A Causa De La Erupción Del Volcán Pacaya Y La Tormenta Tropical Ágatha*. Retrieved October 2, 2019, from https://repositorio.cepal.org/bitstream/handle/11362/26047/S2011022_es.pdf?sequence=1&isAllowed=y

Congreso de la República. (1985). *Constitución Política de la República de Guatemala*. Retrieved October 2, 2019, from https://www.constituteproject.org/constitution/Guatemala_1993.pdf?lang=en

Congress of Guatemala. (2013, September 5). *Framework law to regulate reduction of vulnerability, mandatory adaptation to the effects of climate change, and the mitigation of greenhouse gas effects* (Decree of the Congress 7–2013). Retrieved December 13, 2019, from http://www.lse.ac.uk/GranthamInstitute/wp-content/uploads/2015/05/GUATEMALA.pdf

Copsey, O. (2020). A regional approach to eco-schools in the Western Indian Ocean. In A. Gough, J.C-K. Lee & E.P.K. Tsang (Eds.), Green schools globally: Stories of impact on education for sustainable development. Cham: Springer.

Danish Outdoor Council (2017). *Eco-Schools Best practice in Uganda, Tanzania and Malawi*. http://www.groentflag.dk/media/1737659/doc_best-practice-study_20-june_print-10.pdf

de la Garza, K. (2016). Pedagogical mentorship as an in-service training resource: Perspectives from teachers in Guatemalan rural and Indigenous schools. *Global Education Review, 3*(1), 45–65.

Food and Agriculture Organization of the United Nations. (FAO). (2014). Country fact sheet on food and agriculture policy trends—Guatemala 5.

Gibb, N. (2016). *Getting climate-ready: A guide for schools on climate action*. Paris: UNESCO. Retrieved October 16, 2019, from https://unesdoc.unesco.org/ark:/48223/pf0000246740.

Girvan, C., Conneely, C., & Tangney, B. (2016). Extending experiential learning in teacher professional development. *Teaching and Teacher Education, 58*, 129–139. https://doi.org/10.1016/j.tate.2016.04.009.

Global Facility for Disaster Reduction and Recovery (GFDRR). (2009). *Disaster risk management programs for priority countries*. Washington, DC: The World Bank. https://doi.org/10.1596/978-1-4648-0484-7. Retrieved from https://www.unisdr.org/files/14757_6thCGCountryProgramSummaries1.pdf.

Gonzalez, J. (2016, September 25). How pineapple charts revolutionize professional development. *Cult of Pedagogy*. https://www.cultofpedagogy.com/pineapple-charts/

Gough, A. (2016). Tensions around the teaching of environmental sustainability in schools. In T. Barkatsas & A. Bertram (Eds.), *Global learning in the 21st century* (pp. 83–102). Sense Publishers.

Gough, A. (2020). Transforming education through green schools: Trials, tribulations and tensions. In A. Gough, J. C.-K. Lee, & E. P. K. Tsang (Eds.), *Green schools globally: Stories of impact on education for sustainable development*. Cham: Springer. https://www.springer.com/gp/book/9783030468194

Gough, A., Lee, J. C.-K., & Tsang, E. P. K. (Eds.). (2020a). *Green school globally: Stories of impact on education for sustainable development*. Cham: Springer.

Gough, A., Lee, J. C.-K., & Tsang, E. P. K. (2020b). Green school movements: An introduction. In A. Gough, J. C.-K. Lee, & E. P. K. Tsang (Eds.), *Green schools globally: Stories of impact on education for sustainable development*. Cham: Springer.

Guatemala Sector Brief—Education. (2017, December 1). Retrieved December 13, 2019, from https://www.usaid.gov/guatemala/education

Henderson, K., & Tilbury, D. (2004). *Whole-school approaches to sustainability: An international review of sustainable school programs*. Report Prepared by the Australian Research Institute in Education for Sustainability (ARIES) for The Department of the Environment and Heritage, Australian Government. Retrieved from http://aries.mq.edu.au/projects/whole_school/files/international_review.pdf

Hernández, A. (2012). Cambio climático en Guatemala. Efectos y consecuencias en la niñez y la adolescencia. UNICEF. https://www.unicef.org/guatemala/media/1391/file/Cambio%20clim%C3%A1tico%20en%20Guatemala.pdf.

How FEE EcoCampus works. (n.d.). *Eco schools*. Retrieved April 30, 2020, from https://www.ecoschools.global/how-it-works

Karpudewan, M., & Khan, N. S. M. A. (2017). Experiential-based climate change education: Fostering students' knowledge and motivation towards the environment. *International Research in Geographical and Environmental Education, 26*(3), 207–222.

Karrow, D., & DiGiuseppe, M. (Eds.). (2019). *Environmental and sustainability education in teacher education: Canadian perspectives* (pp. 2–18). Cham: Springer.

Lysgaard, J., Larsen, N. and & Læssøe, J. (2015). Green Flag Eco-Schools and the Challenge of Moving Forward. In V. Thoresen, D. Doyle, J. Klein, & R. Didham (Eds.), *Responsible Living* (pp.135–150). Springer. https://doi.org/10.1007/978-3-319-15305-6_10.

Marlow, M. P., & McLain, B. (2011). Assessing the impacts of experiential learning on teacher classroom practice. *Research in Higher Education Journal, 14*, 1–15.

Ministerio de Ambiente y Recursos Naturales. (MARN). (2017, August). *Política Nacional De Educación Ambiental De Guatemala*. Retrieved October 2, 2019, from http://www.marn.gob.gt/Multimedios/10363.pdf

Ministerio de Educación. (2007). *Curriculum Nacional Base Sexto Grado*. Retrieved December 9, 2019, from http://www.avivara.org/images/CNB_Sexto_Grado-reduced.pdf

Morgen, A., Gericke, N., & Scherp, H.-Å. (2019). Whole school approaches to education for sustainable development: A model that links to school improvements. *Environmental Education Research, 25*(4), 508–531. https://doi.org/10.1080/13504622.2018.1455074.

Oldakowski, R., & Johnson, A. (2018). Combining geography, math, and science to teach climate change and sea level rise. *Journal of Geography, 117*(1), 17–28. https://doi.org/10.1080/00221341.2017.1336249.

Otieno, D., Wandabi, D., & Dixon, L. (2020). Eco-schools Kenya: Practising education for green economy and sustainability. In A. Gough, J. C.-K. Lee, & E. P. K. Tsang (Eds.), *Green schools globally: Stories of impact on education for sustainable development*. Cham: Springer. https://www.springer.com/gp/book/9783030468194

Patrinos, H. A., & Velez, E. (2009). Costs and benefits of bilingual education in Guatemala: A partial analysis. *International Journal of Educational Development, 29*(6), 594–598. https://doi.org/10.1016/j.ijedudev.2009.02.001.

Rainforest Alliance. (2017, September 12). *Getting ahead of climate change in Guatemala's Western highlands*. Retrieved December 12, 2019, from https://www.rainforest-alliance.org/articles/getting-ahead-climate-change-guatemala-western-highlands

Rojas, A. (2011, November 16). *Educación afronta graves deficiencias – Prensa Libre*. Retrieved December 9, 2019, from https://www.prensalibre.com/guatemala/educacion-afronta-graves-deficiencias_0_592140806-html/

Rubio, F., de Véliz, L. R., Mosquera, M. C. P., & López, V. S. (2017). Impact of teachers' practices on students' Reading comprehension growth in Guatemala. *New Directions for Child and Adolescent Development, 2017*(155), 67–76. https://doi.org/10.1002/cad.20191.

Sharma, P. K., & Kanaujia, P. R. (2020). Journey of green schools in India. In A. Gough, J. C.-K. Lee, & E. P. K. Tsang (Eds.), *Green schools globally: Stories of impact on education for sustainable development*. Cham: Springer.

Tal, T. (2020). Green schools in Israel: Multiple rationales and multiple action plans. In A. Gough, J. C.-K. Lee, & E. P. K. Tsang (Eds.), *Green schools globally: Stories of impact on education for sustainable development*. Cham: Springer. https://www.springer.com/gp/book/9783030468194.

UNESCO. (1978). Intergovernmental conference on Environmental Education: Tbilisi (USSR) (Final Report). 14–26 October 1977. Paris: UNESCO.
UNESCO. (2017). *Changing minds, not the climate: The role of education*. https://unesdoc.unesco.org/ark:/48223/pf0000266203.locale=en
United Nations Development Programme (UNDP). (2018). *Human development indices and indicators: 2018 statistical update*. New York: UN. https://doi.org/10.18356/656a3808-en.
Universidad de San Carlos Guatemala. (2009, June). *Rediseño curricular Del Programa de Desarrollo Profesional del Recurso Humano del Ministerio de Educación*. http://www.mineduc.gob.gt/portal/contenido/banners/profesionalizacionDocente/documents/PADEP_D_curriculo_USAC-EFPEM.pdf
USAID. (2017). *Climate change risk profile: Guatemala*. Washington, DC: USAID.
Villegas-Reimers, E. (2003). *Teacher professional development: An international review of the literature*. Paris: International Institute for Educational Planning. Retrieved October 16, 2019, from https://unesdoc.unesco.org/ark:/48223/pf0000133010.
Wals, A. E. J., & Benavot, A. (2017). Can we meet the sustainability challenges? The role of education and lifelong learning. *European Journal of Education, 52*, 404–413. https://doi-org.ezp-prod1.hul.harvard.edu/10.1111/ejed.12250.
World Bank Group. (2018). *Learning to realize education's promise*. World Bank.

Lina Lopez Lalinde is a graduate of the International Education Policy Masters program at the Harvard Graduate School of Education (HGSE). She is a graduate of Northeastern University, where she completed her B.A. in International Affairs and Political Science. Prior to arriving at Harvard, Lina worked at an international school in Panajachel, Guatemala, where she designed and taught Social Studies and English Language Arts curricula to secondary students. She has also worked as a teacher and advisor for university students in Cambodia, Australia, and the United Kingdom, and continues to be passionate about leadership development and facilitating relationships across borders.

Carrie Maierhofer is passionately dedicated to education. She is a graduate of the International Education Policy Masters program at the Harvard Graduate School of Education. She received a B.A. in Elementary Education from Saint Mary's College and a M.S. in Education from University of Dayton. Prior to HGSE she was a classroom teacher for eight years where she became a strong advocate for supporting teachers in both their professional and personal lives. She is committed to ensuring teachers' voices are heard and respected in the development and implementation of curriculum, education policy, and decision making.

Chapter 4
Rezistans Klimatik: Building Climate Change Resilience in Haiti through Educational Radio Programming.

Ashley Bazin and Christelle Saintis

4.1 Introduction

It is almost 5 p.m., in a market in Port-au-Prince. A merchant, noting the time, flips open his phone and places it on top of his *papitas* (plantain chips) and candy basket, waiting for the familiar jingle that announces the beginning of a radio segment. If you observe the current activity in the market, merchants and customers alike are listening to the radio broadcast. At the same time, the chatter in the car of a family of 4 lulls into silence as a parent raises the volume of the radio. The other parent encourages the kids to pay attention, as it is educational, and the 10- and 14-year-old youths listen begrudgingly.

The Port-au-Prince population has been tuning in every day on their radios in order to learn how human behaviors are negatively affecting the environment in their country of Haiti. They are seeking to learn how they can decrease the impact of climate change, as well as how to respond to climate-related disasters. This learning opportunity is especially important for the merchant; since they did not attend school, it is the only way they can educate themselves. In the short time they have been tuning in, the fictional characters in the radio show have become as familiar to them as their own friends and the customers they serve each day. In the episodes, the main characters face critical issues concerning climate change that reflect the local context; the decisions that they make and discussions they have on the program are a reality for many of the people of Haiti. The street merchant and family can relate to the fictional characters and have begun adopting and incorporating some of the suggestions into their lives, as well as discussing the episodes with friends and family.

A. Bazin (✉) · C. Saintis
Harvard Graduate School of Education, Cambridge, MA, USA
e-mail: abazin@gse.harvard.edu

F. M. Reimers (ed.), *Education and Climate Change*, International Explorations in Outdoor and Environmental Education,
https://doi.org/10.1007/978-3-030-57927-2_4

Scenarios such as the two described are part of the daily routines of the Haitian population. While a variety of content is broadcasted via the radio, it is often informational by nature and educational for the listening population. This form of media is even more impactful amongst those who have little to no formal education. Radio programming substantially raises understanding of matters, such as climate change, educating both the formally educated and non-formally educated. The widespread use of radio in Haiti, due to its accessibility, is a factor that we considered when designing a program to help mitigate climate change and build resilience among the population of this island nation. It is undeniable that climate change is one of the most critical issues facing the planet at this moment. Addressing this issue has come to the forefront of international conversations as research and interactions with the environment have rendered the dialogue concerning climate change no longer a choice but a necessity. As it pertains to Haiti, the population experiences the increasing severities and fatalities of climate change every year. There is not a question of their awareness of climate change, however, the goal is to provide a solutions-oriented, comprehensive and science based information of these climatical occurences.

The report, *Education & Climate Change* (Commonwealth Education Hub 2015), acknowledged that mitigating climate change requires collective action as a global community. There is a need for a shift in the global culture, improvement in human-decision making regarding their climatic behavior, as well as an increased awareness of the causal relationship between people and the health of the planet. This is especially important, as more countries become more industrialized, concluding in larger carbon footprints, thus exacerbating climate change and its' impacts - which disproportinately affects developing countries such as Haiti. One meaningful way to address this necessary shift is through education, which can provide agency, inform and motivate all inhabitants of this planet, including local community members and governments (Commonwealth Education Hub 2015).

The issue of climate change is most pressing to developing countries and island-states that are very susceptible to the impact of changes in the environment (UNESCO 2012). One of the most vulnerable countries in the world to climate change is Haiti (Slagle and Rubenstein 2012). This chapter examines how education can help to mitigate climate change in Haiti by building the resilience of the citizens through educational radio programming. We consider climate change resilience in the context of Haiti to be the ability of the people to maintain a certain level of functioning while absorbing the effects imposed upon them by changes in the climate, as well as their capacity to adapt and evolve in ways that improve sustainability, leaving citizens better prepared for the future as impacts of climate change increase (Folke 2006; Nelson et al. 2007).

In this chapter, we explore the following: (a) a comprehensive assessment of current, retrospective and prospective impacts of climate change in Haiti; (b) an analysis of the two-way relationship between climate change and human behavior; (c) an analysis of the use of education to mitigate climate change in Haiti; (d) an examination of the current educational policies and programs currently being implemented globally to mitigate the impact of climate change, informal learning, and how media can develop sustainable practices; (e) the potential implications of climate change education in Haiti as assessed against given criteria; (f) an overview of the modes of media in Haiti; (g) a theory of change for the radio programming; (h) and finally, the implementation plan demonstrating the measures to be executed in order to have an effective program and fulfill the goals.

4.2 Analysis of Current and Anticipated Impacts of Climate Change in Haiti

4.2.1 Geography

Haiti is a tropical island-country located in the hurricane belt. According to the United Nations Development Program (UNDP) adaptation website (n.d.), Haiti has a total land area of 28,000 km^2, and approximately 70% of the island is mountainous, thus the indigenous Taino moniker Ayïti, meaning "land of high mountains" (Ng Cheong-Lum 2005). The country has experienced both natural and man-caused hazards. The current environmental issues include active and extensive deforestation, soil erosion, and overpopulation, which impacts the distribution of potable water. The country is also subject to severe storms from June to October, droughts, flooding/threatening sea-levels, landslides, and earthquakes. Currently, only 3% of the country has forest cover, making way for agriculture and fuel production used for 71% of Haiti's energy requirements (CIA, 2018).

4.2.2 Economy

Though overpopulation is an issue, according to the World Population Review (2020), the country continues to grow at a rate of 1.24% a year, but it will hit its peak in 2070 and then decline. It is the poorest country in the Western Hemisphere, with 80% of the population living in poverty. The livelihood of 70% of Haitians is dependent on agriculture, provided through small-scale subsistence farming. Agriculture is also 32% of the composition of the Gross Domestic Product (UNDP n.d.). Due to its economy and geographic stance, Haiti is very susceptible to the potential damages inflicted by climate change, even more post the disastrous 2010 earthquake.

4.2.3 Other Socioeconomic Factors

Other factors that contribute to the observed lack of resilience to the risks of climate change include government capacity, population growth and density, agricultural dependency, poverty, history of armed conflict, economic sector, public health systems, and weak infrastructure (Singh and Cohen 2014).

4.2.4 Climate Summary – Comparative Approach

According to the report *Climate Risk Profile: Haiti* (USAID 2017), the climate trends in Haiti since 1960 include rising average annual temperatures of 0.45 ° C, most of which occur during the warmer months of the year, from June until November. There was also a rise in the number of hot days per year by 63 days, between 1960 and 2003. The average rainfall has decreased each decade by 5 mm. Natural hazards, such as hurricanes, have not only grown in numbers but also substantially increased in intensity. Many of the low plain areas and coastal zones have also experienced more flooding.

The USAID report also predicted some of the trends expected in Haiti in the future, including average annual temperatures rising between 0.78° and 2.16 ° C by the year 2050. Annual precipitation will continue to decrease by up to 43 mm, with droughts being a common re-occurrence, thereby causing problems for agriculture and farmers. Despite the predicted decrease of annual rainfall, there is also the possibility that the rainfall may become more extreme. Hurricanes, which have devastated the country, will grow more intense by between 5–10%. The surrounding sea level is also expected to rise by up to 0.4 m by the year 2030 (Climatelinks 2017).

4.2.5 Future Impacts

Many different sectors and industries stand to be affected by changes in the climate. Indicated below are several anticipated impacts that will occur if Haiti does not address climate change.

4.2.5.1 Agriculture

In the future, climate change will most likely impact farming and agricultural export operations, which contribute significantly to Haiti's GDP. Furthermore, if rainfall declines and the temperature continues to rise, crop yields will be affected and negatively modify the country's nutrition, as well as families' ability to earn and children's ability to go to school. Also, the impact on the viability of agriculture will result in a decrease in Haiti's competitive edge on crops such as coffee, cacao, and mango in the international market (Smucker et al. 2007).

4.2.5.2 Health

Climate change will significantly worsen the already weak health system in Haiti. The stressors imposed by climate change will increase the disease burdens and exacerbate existing health issues, such as outbreaks of both vector and waterborne diseases. Natural hazards destroy road infrastructures decreasing accessibility to health services. The poor will suffer even more from an increase in malnutrition (Climatelinks 2017).

4.2.5.3 Education

Overall, damages to the road infrastructures will cripple the ability of students and teachers to travel to and from school. Also, school closings will be more common due to the instability of the buildings or their use as emergency shelters. Economic hardships, food security issues, and other factors impact school performance and attendance, leading to a regression in the quality of and access to the education system (Climatelinks 2017).

Beyond the external impacts on the educational experience, Haiti already has a feeble education system. A retroactive analysis of the education system that concluded in the adults of today is as follows. During the Duvalier era (1960–1980), one ran by a Father-Son dictatorial duo, there was a decline in the quality of education in Haiti because many qualified teachers fled the country due to political tension (Salmi 2000). According to Hadjadj (2000) – in the 1980s and 1990s, although there was a significant increase in the number of teachers, the quality of these teachers left much to be desired. Only 11.3% of primary school teachers held a professional degree (defined as teacher training school graduates and holders of teaching diplomas) in the 1980s, and 15% in the 1990s. Also, there was a significant number of "over-aged" students with, 53.2% of students being three or more years behind in relation to the legal age in 1980 and 48% in 1996.

4.2.5.4 Economic Factors

Climate change can result in the reduction of revenues, damages to a multitude of infrastructures (i.e., transportation and export), and a decrease in investor interest. Furthermore, it will involve the diversion of funds to cover the costs of damages caused by natural hazards. The exacerbation of the fragile economic sector will result in the entire country enduring the stressors imposed on them (Climatelinks 2017).

4.2.5.5 Possible Domestic Solutions

Given all these elements, in order to increase climate change resilience and decrease the negative outcomes of climate related disasters, Haitians need to adopt sustainable agricultural methods, engage in healthier volumes and types of consumption, and set forth policies that protect their part of Earth.

4.3 Connecting Climate Change to Human Behavior

According to the scientists studying the causal relationship between humans and the environment, the current and problematic state of the climate is directly related to the choices and behavior of humans (Psychology and Global Climate Change 2009). Actions such as increasing population and consumption patterns have worked together to influence the changes in the environment. Individual predictors of consumption include income and skills, connection to nature, perceptions of needs versus luxuries, core psychological needs, etc. In turn, the implications of climate change for humans cannot be understated. Increasing levels of stress on the environment will inevitably lead to resources becoming scarce. At a psychosocial level, the scarcity in resources and preexisting inequalities will affect interpersonal and intergroup behaviors, possibly concluding in stress and anxiety. When considering this in the face of already existing competition and gaps in the wealth of countries and individuals, the effects will undoubtedly be profound. A potential positive consequence of climate change is international unity that can lead to us working collectively to mitigate the changes (American Psychological Association Task Force 2009) Indeed, acknowledging that the effects of climate change are not restricted to specific countries or regions but instead are felt worldwide, emphasizes this need for international collaboration.

Williamson, Satre-Meloy, Velasco, and Green (2018), made a case for a range of behavioral approaches to climate change, and acknowledge that individual behavioral changes, if on a large scale, can make a significant impact. They identified a total of 30 behavioral changes that can mitigate between 20–37% of global emissions for the next 30 years. Regarding human contributions to global warming, it has been occurring since the mid-twentieth century, via greenhouse gas emissions which are activated by decomposition, fossil fuel burning, and chemical fertilizers. Essentially, the cause of climate change is the imbalance between human-made deposits into the natural systems and the speed at which the natural system processes them. What is needed is an overall decrease in the consumption of raw natural resources by humans, specifically those of high-income countries. Additionally, they recommend: reduction in food waste, plant-rich diets, more efficient energy use, planting more trees, farmland restoration, conservative agriculture, reducing the use of fertilizer, installing irrigation systems, electric vehicles, mass transit, bicycles, solar rooftops, LED lighting, household water-saving and a slew of other behavioral solution to reduce our ecological carbon footprint.

Evans' (2019) research on Projected Behavioral Impacts of Global Climate Change is also relevant. He argues that global climate change can affect human behavior, including increased intergroup conflict (uptick in aggression and crimes), adverse mental health consequences (anxiety, suicide, etc.), deficits in cognitive developments, and Post-Traumatic Stress Disorder. Climate change alters the quality of life, which is foundational in mental health outcomes. The tipping point can be as simple as the restriction of outdoor recreational play resulting in psychological distress.

Psychological approaches can also be deployed to modify human behaviors in ways that can reduce and potentially reverse the impact of climate change. Van Der Linden et al. (2015) described five best practices for improving public engagement with climate change. First, Social, Cognitive, and Clinical psychology have demonstrated that the human brain prioritizes direct personal experience over abstract occurrences. While data is necessary, in the effort of encouraging human behavior, it is ideal also to highlight climate change-related changes in the different ways a human experience it. Second, humans adhere to social and group norms. Humans, as a collective, influenced the climate changes we are currently witnessing; in order to change it for the better, it will require group effort. Therefore, concerning climate change, the sense of responsibility at the individual level is incredibly low. In the face of this issue, policymakers should try to stimulate positive behavior by re-establishing good prescriptive (how to behave) and descriptive (how others behave) norms. Third, there is a lack of urgency amongst humans; when making decisions, and evaluating costs and benefits, people discount uncertain future events. Due to the hazy knowledge of the future impacts of climate change, the consequences of climate change feel psychologically and physically far from the current moment for most people. One way to cultivate urgency is by bringing forth the current manifestations of climate change, explicitly recognizing regionally relevant impacts. Fourth, Van der Linden et al. argue that when it comes to climate change, people are prompted to take action when climate change losses are framed as a certainty that occurs in the present versus losses paired with uncertainty in the future. In response to that evidence, the conversation around climate change should be focused on the positive gains of immediate actions versus the potential losses of the future due to climate change. Lastly, to spur on a change in human behavior, human motivation should be used as a resource. Both intrinsic (caring for others) and extrinsic (money) motivations should be utilized. It is best to cater to both intrinsic and extrinsic motivations together because it is more sustainable, mainly due to climate change being a long-term environmental problem.

4.4 Exploring Different Alternatives of Delivering Climate Change Education

4.4.1 The Use of Education in Addressing Climate Change

It is essential in this analysis to understand what the existing literature states on education's role in combatting climate change. The 2015 United Nations' Sustainable Development Goals (SDGs) explicitly state the importance of education in helping to reach the goals by 2030 (UNESCO 2015). Many organizations have incorporated this urgency of education at the center of climate change action. The goals of these organizations are all related to the emphasis they place on the importance of climate action and the role education can play.

The National Education Association (NEA), one of the two major national teacher unions in the United States, echoes the importance of educating society on the impact of climate change and global warming that are found in the SDGs (n.d.). They explain that teaching people the essential message that climate change not only exists but has implications and consequences for all humans and the Earth we share is paramount. The National Oceanic and Atmospheric Association (NOAA) has also laid out several key facts all students should know and understand, including that the effects of climate change can already be seen, that current infrastructures are in danger, and that climate change will likely threaten agriculture and thereby food security in the future (NOAA n.d.). All these factors are of great importance in the context of Haiti.

Curtailing the effects of this global crisis will involve all members of society – not just politicians or those with access to information – and, therefore, will require a transfer of knowledge. In addition, learning from education is necessary because it is critical to address these issues immediately instead of only learning from experience (Ledley et al. 2017). Adequate and quality climate change education should emphasize the ability to achieve positive outcomes with activities framed around the gains that can be made by reducing individual footprints as well as the use of persuasive language to incite changes in behavior.

Addressing climate change through education is of global importance and is especially salient for the island nation of Haiti. If the country is to mitigate these effects, they must turn to education and recognize the crucial role it can play in helping to launch climate action in the country (Commonwealth Education Hub 2015). Indeed, education can support social change and significantly influence climate change mitigation, by increasing the number of citizens making informed decisions, regarding potentially destructive habits (Commonwealth Education Hub 2015). As stated by UNESCO, climate change education can help "people understand and address the impact of global warming, encourages changes in their attitudes and behavior and helps them adapt to climate change-related trends" (UNESCO n.d., para.1, 2010). Promoting a deeper understanding and awareness of the problems the planet and local communities are facing can help to shift public attitudes and ways of thinking on this topic. Education is also an essential part of helping individuals and communities learn to adapt to trends and empowers them to adopt lifestyles that are more sustainable (UNESCO 2010).

4.4.2 Current Climate Change Curricula; Best Practices and Common Themes

Most educational approaches to climate change involve work in formal education institutions. As important as working with children in school is, not all children attend school, and many adults have been in school for very few years. This reality calls for the development of non-formal and informal approaches to climate change education.

There are several institutions that have begun to create a climate change education curriculum, including The National Education Association (NEA), The National Oceanic and Atmospheric Association (NOAA), Stanford University, and UNESCO. In addition, several countries, such as China and Singapore, have begun to incorporate climate change curriculum into their education systems (Læssøe et al. 2009). In reading the literature, we were able to identify the successes and failures of different climate mitigation via education projects. Below is a list of the commonalities shared amongst the curriculum implemented around the world.

1. The integration of climate change into the school curriculum has been formulated to be a gradual and continuous mode of learning throughout the academic years. (Climate Curriculum n.d.)
2. In practice, many have integrated current data on climate systems into their STEM courses, including subjects such as Geography, Physics, Chemistry, etc. (Stanford Earth n.d.).
3. Others have opted for a more interdisciplinary approach, given the different ways in which climate change impacts people. For example, in S. Korea, the Education for Sustainable Development initiative included three perspectives: socio-cultural, environmental and economic based on the on-going realities of climate change in the country (Læssøe et al. 2009)
4. Another more pragmatic approach is Climate Change Education for Sustainable Development, which is an initiative promoted by UNESCO in developing countries such as the Dominican Republic. This type of education is focused on building local capacity through formal, non-formal, and informal projects. These projects in local communities are run by government agencies, civil society organizations, and empowered young leaders (UNESCO 2019).

4.4.3 Government Action

Despite the attempts to date, to combat the effects of climate change, Haiti's institutions still face numerous challenges at both the national and local levels. The scarcity in financial resources, the lack of technical capacity, the unreliable energy infrastructure, and the political instability prove to be obstacles in the process of allocating resources, and the implementation of different programs during times of disaster recovery. The attempts at achieving sustainable growth and development impacts the government's ability to provide essential services. These struggles exist during the regular occurrences of natural hazards and will continue to worsen with the incoming effects of climate change (Climatelinks 2017).

For example, the Dominican Republic piloted UNESCO's Climate Change Education for Sustainable Development (ESD) Program. This is especially pertinent given that the Dominican Republic and Haiti share the same island. In the

Dominican Republic, climate change education has gained momentum since its inclusion in the Constitution of the country in 2010. In schools, ESD is taught in an interdisciplinary way using a variety of pedagogical practices in order to promote participatory learning. Training for teachers has been made available and includes support on building climate change into the curriculum as well as a research program that helps to promote sustainable development. Climate change education is focused on building local capacity through both formal, non-formal, and informal projects. These projects in local communities are run by government agencies, civil society organizations, and encourage the development of young leaders (UNESCO 2015). While these projects seem promising and could be applied in Haiti, financial constraints and corruption are some of the limitations of implementing these strategies (Gallagher et al. 2019). In addition, only 29% of Haitians 25-and-older have attended secondary school, and approximately 50% of children do not attend school (Haiti Partners 2018). Due to these constraints and goals, the use of non-formal and informal means of delivering climate change information and education is more relevant in the context of Haiti.

4.4.4 Informal Education

Informal education encompasses any teachings outside the systematic organization of formal education. The typical characteristics of informal education include a decrease in contact between teacher and student, as well as learning taking place outside of an institution. Additionally, it is flexible, and the methodology varies in structure. This type of education does well in efficiently and effectively meeting the needs of a society because it is very functional and practical. Informal Education can be instrumental in developing countries due to its innate use of multidisciplinary strategies that are compatible with existing realities to mediate quandaries (Dib 1988).

In the context of Haiti, informal education is essential to reach the adult population. Working through the formal education system would prove ineffective for several reasons. It would further strain an already weak infrastructure, and with an existing scarcity of teachers and resources. Attempting to introduce climate change education formally may not reach as broad an audience and have as large an impact as informal education, especially when considering the high number of people out of school, or that have little-to-no formal education. Also, formal education as it pertains to climate change is essentially a long-term investment form of climate action. Due to the existing and dire circumstances of climate change, measures to be taken to mitigate the effects of it need to be quick and swift – therefore, using informal education can prove to be effective in rapidly accomplishing this.

4.4.5 *The Use of Radio in Sustainable & Community Development*

Authors, researchers, and activists have identified radio as a powerful and cost-effective way of addressing issues of sustainability (Chapman et al. 2003; CIMA 2007; Jayaprakash and Shoesmith 2007; Khan et al. 2017; Sewlal 2014). There are several examples from different countries on the positive impact that radio has had on the communities and the spread of knowledge and information. Radio is more effective than other forms of media communication due in part to the low costs, access, reach, and portability (Jayaprakash and Shoesmith 2007).

In Ghana, for example, radio was used to improve the sharing of agricultural information in remote farming villages (Chapman et al. 2003). In Chapman et al.'s (2003) study, radio programs, which were broadcasted in the local languages, used drama performed by local actors along with discussions relating to the changes desired in the behaviors of the farmers, in order to determine if the radio show could influence positive behavioral change. The researchers chose the radio format to deliver the messages because of the strength of that media tool in reaching a broad audience, most of whom were illiterate, and provide them with the necessary information on agricultural production in a way that they understood. This meant using the local languages but also relaying the information in the local communities' agricultural vernacular, required understanding the farmers' knowledge of the subject before creating the program itself. The agricultural radio program in Ghana was successful in changing the behaviors of 94% of the farmers to behaviors that were more sustainable. Other direct results of the initiative were public debates on important matters, the introduction of new agricultural techniques, changes in public perceptions, as well as a balance between indigenous and scientific practices regarding agriculture. In this way, radio was able to address a more significant complex social problem, something that other countries have done as well.

In South Africa, the Soul City Project uses radio and other forms of media to address problems such as violence against women, HIV and AIDS, as well as other health issues (Chapmen et al. 2003). The educational messages delivered on the radio for the project are based around the process of 'edutainment,' which encourages debates and discussion on difficult and controversial topics, rather than simply dictating a message. The country uses the radio to give exposure to local issues, and broadcasting has helped with the development of small businesses, discussions on gender equity, consumerism, sports, music, and climate change (Sewlal 2014). By airing lessons and other forms of news and entertainment, these radio programs have been able to make tangible differences in the communities.

Some of the limitations of radio in sustainable and community development lie in a lack of funding, trained personnel, and poor leadership (Sewlal 2014). In many of the examples, the participatory nature of radio could be on the one hand empowering for localities while, on the other hand, challenging in terms of staffing capacity: since many stations rely on volunteers, problems can arise because of a lack of talent, time commitments, and high turnover rates (Sewlal 2014). Also, the operating and maintenance expenses can cut deeply into already thin funding.

Despite these limitations, radio remains the most accessible form of mass media and can capture broad audiences that are both literate as well as illiterate (Khan et al. 2017). It enlightens communities and can educate them on initiatives and strategies that will result in improvement in the lives of its listeners on issues surrounding agriculture, health, and environmental degradation, and more. It can also provide a lifeline to local communities in times of natural hazards (Sewlal 2014) and replace fear with hope through the sharing of information (CIMA 2007).

4.5 Review of Current Education Policies and Programs to Address Climate Change in Haiti

As previously explained, the importance of education in the fight against climate change cannot be understated. In response to global changes, Haiti must invest in developing and implementing comprehensive educational programs for society by addressing the concerns and solutions for climate change. There is existing research on the measures of resilience recommended for the government to consider when designing the efforts to rebuild after the earthquake (Singh and Cohen 2014). However, many of these efforts currently do not include a way to incorporate climate change in educational programs. Instead, many international organizations, such as USAID, have a focus on increasing government capacity and aiding the Ministry of the Environment. They also work to lead efforts in cutting back on behaviors that result in an increase in the effects of climate change (UNDP n.d.).

It is important to note that a necessity for the implementation of climate change programming in Haiti is funding. Financial constraints continue to be a substantial prohibitive factor in the country's efforts to combat climate change. Gallagher et al. (2019), analyzed the allocation of Haiti's financial assistance dedicated to Climate Change projects. They observed that this funding is dedicated to four of Haiti's 23 priorities for building resilience against climate change: food security, disaster risk reduction, renewable energy, and integrated water resource management. However, information and education are included in the list as a priority for the country, which highlights Haiti's recognition of education as a crucial part of its sustainable development and fights against climate change (Gallagher et al. 2019).

They also noted that some of the finances had been diverted to disaster recovery caused by regular environmental stressors, such as hurricanes. Otherwise, other matters such as agricultural adaptation, afforestation, coastal-zone management, capacity building, development of bioeconomy, etc. are relatively neglected. The analysis recorded a total of 76 climate change funded projects. Some date back to 6 years ago, but most were initiated after 2015. Of these 76 projects, half were categorized as mitigation projects while the other half were considered adaptation projects. It is worth noting that, though active, there is no verification of the effectiveness of these programs. Gallagher et al. (2019) alluded to the lack of institutional regulations to combat climate change. They made recommendations regarding creating effective policies targeting different sectors, culminating in the creation of a comprehensive institutional environment that would be more mindful of the allocation of funding and reduce the need to manage all the climate change projects. However, a focus on education is not mentioned in their report (Gallagher et al. 2019).

Overall, Haiti is being reactive to climate change as opposed to proactive. As previously mentioned, most of the programming related to climate change in Haiti is either mitigating in nature or adaptive in approach. The lack of funding poses as a prominent obstacle in the goal of building resilience against climate change. In order to have a lasting effect, the country will have to adopt policies and programs that reach a broad audience, and this needs to have a two-tiered focus. Therefore, some policies and programming should focus on preventative measures of climate change, and others will have to target behaviors and choices at the community and individual level.

4.6 Implications of Climate Change Education in Haiti

Considering its vulnerability to climate change, the importance of educating the population in Haiti on how to mitigate these effects, as well as the impact on the local environment, cannot be understated. From the existing literature (Læssøe et al. 2009), there were several themes and best practices deemed most useful and relevant to the island nation's future climate action.

The first is the recurring theme of climate change education as a way to build skills. In Haiti, there is a critical need to develop the population's capacity to adapt to the changing environment. The population needs to learn how to make choices that are sustainable. For instance, what are the ways in which Haitians survive through. The ever-increasing number of severe natural hazards, such as hurricanes, to decrease the number of direct and related fatalities?

Second, learning that is related to the local environment and context is also essential. In some examples and cases found in the literature, the content of climate change initiatives is directly related to the context in which people live instead of including broad overarching knowledge about the phenomenon in general. This should work exceptionally well in Haiti; while Haitians are aware of climate change, given that they live through the increasing severities of natural hazards, being able to identify the intricate, comprehensive and scientific details of the effects that it has on their local environment would be very beneficial, in order to aid in the development of strategies to address and reduce the impact and limit the risk of future changes.

Third, in order to gain the skills necessary to survive, adapt, and minimize impact, an approach that reaches a relatively broad audience of people should also prove useful. In 2017, USAID reported that the majority of Haitians lack access to quality education. There are several challenges that Haiti faces concerning education. These include low school enrollments, poor literacy rates, lack of government oversight, and shortage of qualified instructors. Any program that wishes to make changes to the behavior of the general population should not focus on the school curriculum. Instead, a program that has widespread access to people who may not have the opportunity to benefit from a quality education makes the most sense.

In order to reach a broad audience, we must consider the media that is available. In the case of Haiti, the radio is the most popular form of media consumed in the country. Television (TV) is solely popular amongst folks who have access to electricity. Interestingly enough, radio stations have their corresponding TV Stations. Additionally, the average Haitian watches TV 4–6 h per week, and the peak time for watching is between 7 pm and 10 pm (Internews 2013).

Another form of media is newspapers, all of which are published in French in Haiti. In addition to the language it is written in and considering that newspapers do not rival radio in terms of popularity, the newspapers are a privilege of the elites; they are not accessible by at least half of the adult population due to lack of literacy in the French language. This relates to physical newspapers; though data concerning digital newspapers are not mentioned here, it is safe to infer that just as for TV, if there is no constant stream of electricity, then there is no Wi-Fi or computer, a reality for much of the country (Internews 2013).

Given these facts in the context of Haiti and the criteria of (1) wanting to reach a widespread audience, (2) half of the population not attending school, (3) many adults not being literate, and (4) a lack of televisions and a widespread use of radio, we identified educational radio programing as having the potential for the most impact in changing behaviors that lead to climate change resilience.

4.7 Overview of Media in Haiti

4.7.1 Radio Use and Stats

In Haiti, the radio is the most popular source of news and information (Internews 2013). This is due in part to few Haitians being able to afford to buy newspapers, and only half of the adult population being able to read and write. Many radio stations quote from the newspapers themselves, making the radio accessible and relevant to the broader community who cannot afford to buy newspapers or who cannot read.

According to the *Media and Telecoms Landscape Guide* (Internews 2013), almost every household in the country has a radio set, and, with the advent of the cell phone, many people now listen to the radio on their mobile devices as well. With more than 4.5 million people owning mobile handsets, the radio is very accessible to the broader population. In a media consumption survey on behalf of USAID, the local market research firm DAGMAR discovered that 96% of the people in the capital city of Port-au-Prince listen to the radio every day (Internews 2013). The survey also revealed high radio listening rates in nearby towns as well as in more distant villages.

Most of the people in Haiti listen to the radio for the news, with music, religious programs, and sports making up a smaller percentage of the reasons for listening. The survey showed that the preferred source of information was the radio, according to 83.7% of the respondents. Electricity is very unreliable in Haiti, making other media such as television more limited, with only 7% of people in the survey saying they preferred this source of media for the news and information; less than a third of Haitians have access to a TV at all.

4.7.2 *Radio and Learning in Haiti*

Since radio is the most accessible and most widely used media source in Haiti, much of the population who are not literate depend on it for their information. In the past, radio stations have informed listeners on educational topics, such as geography lessons and various catastrophes and consequences (Roca 2016). Locals have reported that the radio also serves to inform them on how to behave in society. For those who did not attend school and cannot read the newspapers, about half of the population, the radio can be one way that they learn. Many individuals tune in using their mobile phones, which allows them to access the radio from anywhere. This makes the radio a powerful tool in reaching the wider population and inciting change.

4.8 Theory of Change

The Theory of Change for adopting radio programing to educate for climate change is as follows: If the adult population in Port-au-Prince (1) listens to an educational radio program discussing specific behaviors, such as safety practices that should be taken during natural hazards and (2) changes their behaviors as a result of the radio program, then the resilience to climate change will increase among the population of adults in Port-au-Prince as people engage in behaviors that reduce the negative impact on their lives. We propose to implement this radio program in the capital as this is the area with the largest concentration of people as well as the most widespread use of radio. There are several different behaviors that the radio program could potentially focus on, such as responses to natural hazards, safety practices, and even sustainable agricultural practices.

4.8.1 Audience and Impacts of Project

Many sustainable development educational programs target children. Though necessary, due to the dire circumstances in Haiti and the need to respond immediately to climate change-related disasters, we believe it would be best to target the adult population in order to spur on immediate changes in behaviors and build resilience. Furthermore, given that Port-au-Prince is the biggest city of the country, and houses 2.3 million of the 11 million citizens, we believe that this is the ideal location to target with the start of this project (World Population Review 2020).

Our goal is to address climate change by improving decision-making in order to increase resilience; therefore, we must consider the fact that those making most of the decisions in the adult population are not in school and have not been to school. As a result, reaching this population of adults will be the most effective using radio since only 7% of the population have televisions.

There are several domains that we see our project impacting. These domains are based on a framework for evaluating the impacts of informal science education projects (Friedman 2008). The domains this project will impact are (1) awareness, knowledge or understanding, (2) behavior, and (3) skills. Awareness, knowledge, or understanding refers to an impact that includes what the population in Haiti knows, in their own words, about the safety measures that should be taken during natural hazards. This knowledge should be determined before the start of the programming in order to ascertain any misconceptions that exist, and that can be addressed through the radio program. The knowledge impact category also refers to what people learn during, immediately after, or even long after the listening experience. Behavior refers to measurable changes in the exercise of behavior as related to the topic of safety measures and practices. Skills refer to any impact on the development or reinforcement of skills, in this case, new safety skills for responding to a natural hazard. Therefore, the definition of the best outcome for this project is defined as a shift in culture and behavior, prompting a productive response from the population to climate change issues (Radical Partners 2019).

Other concrete ways we plan on enabling the best outcomes include: designing a curriculum that is context-based and based on safety measures to be taken in response to climate change in Port-Au-Prince; producing a show based on the curriculum as mentioned earlier that is linguistically inclusive; pitch this initiative to the most-listened-to radio station which ensures widespread access; building on the sustainability of this project/show by maintaining good relationships with all stakeholders; and finally creating strict protocols on conduct, length, and material concerning the show.

4.8.2 Measuring Outcomes

The intended goals and objectives of the radio program are improving knowledge surrounding safety practices, changing behavior, and developing the skills necessary to respond to natural hazards. In order to determine whether these goals were accomplished there is certain evidence we would have to collect to support our assertion that the program worked. Concerning the measurement of outcomes, the baseline would be the documentation of past climate change-induced disasters, specifically hurricanes. Every year between June and October, Port-Au-Prince experiences tropical storms/hurricanes. Due to climate change, these tropical storms have been increasingly more disastrous over the years. The outcome measurement would be a comparison of adaptive and recovery behaviors between post-earthquake storms and the storm that occurs after the broadcasting of this show. The analysis would question how people have tried to build resilience pre-storm and how they have tried to recover after the storm. It would also determine if there are differences between the magnitude of the aftermath pre-and-post the show. We would also incorporate a feedback system that would give us insight into how receptive the population is to our show and what their interests are.

Evidence for the impact on knowledge may come from changes in the listener's awareness and understanding of safety measures, which may be directly assessed through observations/surveys or self-reported. Evidence for changes in behavior might include speaking with residents in Port-au-Prince to determine any self-reported intentions to change their behavior, or even follow-ups longitudinally to see if changes in behavior have indeed occurred. Follow-ups might be particularly important to see actual behavioral changes. Finally, evidence for skills might include reflections from listeners or direct observations of the development or practice of new skills promoted in the radio program. Data can simply be gathered through Baseline surveys, which shall then be followed-up by either midterm or end-line surveys. We can try and gain insight into how useful they find the content to be, as well as what new practices they have engaged themselves in. Another form of data is utilizing third-party NGO research as insight into how the area has improved regarding climate change behavior.

4.9 Implementation Plan/Curriculum

4.9.1 Stakeholders

In assessing the implementation of this project, we worked on accounting for different stakeholders in order to identify how we may efficiently and effectively put this project into action. The first stakeholder is the audience, who are on the receiving end of this broadcast and whose behavior we hope to impact. We will try to broadcast our show on a popular radio station; therefore, the network who is willing to host our project is another stakeholder. Domestic climate change-related actions are typically launched by the government via different policies. There is a possibility of building a relationship with government personnel in order to tackle this problem together and devising how to change the culture. All other stakeholders represent the different moving pieces of this project, and this includes actors, writers, scientists, directors, producers, radio managers, financial managers, technical managers, programming manager, etc.

4.9.2 Goal of Program

The goal of the radio program is to bring awareness and influence the behavior of Haitians living in Port-au-Prince in order to increase their resiliency to the changes in the environment due to climate change. One huge impact of this is the growing number of hurricanes and natural hazards, which have plagued Haiti annually and have increasingly inflicted more damage to the people's quality of life. Ideally, we see this radio program as one part of a larger ongoing project that will eventually address a broad range of skills and behaviors that adults need to know in order to increase their resilience to climate change. Given the possible domestic solutions described previously, we envision these skills as blocks that would build on each other over time. While addressing safety practices during natural hazards is a priority and immediate concern, this is just a proposed starting point justified by the demographics of this specific context, as explained in the *Audience and Impacts of Project* section of this chapter.

4.9.3 Topics Covered & Objectives

In our radio program, we will be addressing the concern of safety practices during natural hazards as the first in a series of topics. Being context-specific, the most recurring natural hazard is hurricane season; therefore, we will broach the topics of hurricane preparedness, measures to be taken during the hurricane, and how to recover post-hurricane all within the 'safety' frame. For example, one safety measure that the radio program will cover for pre-natural hazard is crafting an emergency plan that explains how to plan and protect yourself and family, shelter locations, and safety plans for those who live by the water. Post the natural hazard; listeners will learn what to avoid, such as power lines, damaged buildings, floodwaters, and wet electrical devices. A complete list of proposed topics to cover is in Appendix A.

Given that Port-Au-Prince is a metropolitan city, the most eminent impact of climate change has been Hurricane Season. However, a fundamental aspect of this project design is the accommodative and inclusive nature to all climate change-related topics, as a significant piece is to be context-specific. There are broader skills and behaviors that Haitians need to learn in order to effectively build climate change resilience. For example, proper disposal of human waste and trash, how to preserve natural resources, as well as sustainable farming practices. A list of proposed topics that can be covered is in Appendix B.

4.9.4 Strategy

We propose a radio show that consists of ten 10-min short dramas. Table 4.1 shows the anticipated impact of radio shorts on behavior. The fictional series will broadcast several times a day on a commercial radio station that has a large following in Port-au-Prince. The 10-minute dramas will be integrated into the local news broadcast, preferably in the most popular radio station, which in the context of Haiti is *Radio Caraibe.*

The program will be broadcasted pre-, during, and post-hurricane season, which occurs every year between July and October. Out host radio station will play the educational drama series at different times during the day, modeling the frequency of commercials. We believe a repetitive cycle of broadcasting will have the highest impact, as displayed by previous efforts of radio programming (Chapman et al. 2003).

Table 4.1 Impact of radio shorts on behavior

Impact	Impact Category	Audience Objective	Evidence
The listener will change their behavior or acquire new behaviors that increase their safety habits and practices during natural hazards	Behavior	Listeners will have an increase in awareness of the damages to be anticipated and be given tools to best prepare for it.	Pre- & Post-survey. This survey will act both as evidence of behavioral changes, as well as feedback from the audience. We believe that a part of the sustainability of the program lies in the consistent evaluation of the effectiveness of this program. Therefore, feedback is essential for this process. This could be done through baseline surveys followed up by either a mid-term or end-line survey.
		During the occurrence of the hurricane, listeners will participate in safety practices to decrease the amount of harm.	
		Post-hurricane, listeners will gain the skills to easily adapt and adjust to the changes that have occurred and employ problem resolution.	

Furthermore, referencing the reciprocal relationship between Climate Change and human behavior, in regard to content, we do intend to relay the information in a manner that prompts behavior by (1) framing climate change losses as a certainty that occurs in the present versus losses paired with uncertainty in the future, (2) activating human motivation by utilizing both intrinsic (caring for others) and extrinsic (money) motivation.

4.9.5 Pathway for Delivery

Ideally, we will propose this show to the top radio station, which is *Radio Caraïbe*, but others such as *Radio Métropole,* and *Radio Lumière,* are very popular.

4.10 Discussion

The implementation of this project requires effective leadership, and trained personnel in matters such as technical management, creative staff, programming, and financing. Haiti may be deficient in these resources, and we may have to take our involvement a step further and train individuals because a shortage in any of these areas is a threat to the sustainability of the radio show (Jayaprakash and Shoesmith 2007).

Although the international community has taken steps to address the issue of climate change, there is still a great deal of work to be done in the development and implementation of climate change education. As a nation at very high-risk for being affected by these changes, it is imperative that Haiti begins to address ways in which it can adapt to the already rapidly changing environment, as well as how it can strive towards mitigating the future impact of this global phenomenon. Based on the research evidence on the power of education in changing human behavior, which in turn minimizes environmental impact, education in sustainable development is recommended for the country.

Much of the information available on current climate change education shared overarching themes, without many examples of practical ways to best influence students and wider populations. Though we determined best practices in the available cases, the context of Haiti makes it difficult to adopt many of these. In order to prove successful and gain momentum, climate change education in Haiti must be approached in a systemic and multi-sectorial way that includes the common principle of participation and is focused on practical applications through a strengthening of resilience by building skills and capacities.

Our suggestion after reviewing the literature on informal education and the use of media in disseminating information, was for Haiti to approach mitigating climate change through an educational radio program. In this way, we believe we can reach a large audience and truly have an impact on changing behaviors towards those that are more sustainable. As the first step in a continuing series, we are looking at how people can better prepare for naturally occurring events, such as hurricanes, which can be devastating. Through educational radio programming, already the most popular way for people to get information, we believe that climate change resilience among the Haitian people can be increased for the better.

Appendixes

Appendix A

Pre-Natural Hazards – Safety Measures

1. The Difference between Hurricane Watch vs. Hurricane Warning
2. Putting together a basic disaster supplies kit: Basics, Additional, Maintenance and Storage Locations
3. Crafting an Emergency Plan: How to plan and Protect yourself and Family, Shelter locations, how to care for your pets, and safety plans when you live by the water.
4. Clean Up

During Natural Hazards – Safety Measures

5. Power Outage
6. Q&A Session

Post-Natural Hazards – Safety Measures

7. What to Avoid: Power Lines, Damaged Buildings, Floodwaters, Wet electrical devices
8. What to do if you have been in floodwaters.
9. How to Prevent Carbon Monoxide Poisoning
10. Food and Water
11. Taking care of Emotional Health

Appendix B

Other - Topics

- How to preserve your natural resources: A focus on the decrease in the erosion of the land by humans and natural hazards.
- How to engage in industrial behaviors without generating extreme pollution and chemical exposure.
- How to properly dispose of human waste

- How to properly dispose of trash
- For the farmer – how to maximize reaping what you sow, all while practicing preservative methods of farming the land
- Better Eating Habits: Plant-Based Lifestyle
- How can your family engage in Climate-Change Behavior?
- A discussion: What policies should as a collective implement for best practices.
- Dealing with anxiety as it pertains to Climate Change
- How to Prevent, Deal and Recuperate … a PSA:
  - Floods
  - Droughts
  - Tsunami
  - Volcano
  - Storms (Blizzard, Ice, Hail) Tornado
  - Earthquake
  - Wildfires
  - Landslides

References

American Psychological Associations Task Force on the Interface Between Psychology and Global Climate Change. (2009). *Psychology and global climate change: Addressing a multi-faceted phenomenon and set of challenges*. Retrieved from https://www.apa.org/science/about/publications/climate-change-booklet.pdf

Center for International Media Assistance (CIMA). (2007). *Community radio: Its impact and challenges to its development* (Working group report). Washington, DC: CIMA. Retrieved from https://www.cima.ned.org/wp-content/uploads/2015/02/CIMA-Community_Radio-Working_Group_Report_0.pdf.

Chapman, R., Blench, R., Kranjac-Berisavljevic, G., & Zakariah, A. (2003). *Rural radio in agricultural extension: The example of vernacular radio programmes on soil and water conservation in N. Ghana* (Network paper no.127). London: Agricultural Research and Extension Network. Retrieved from https://www.odi.org/publications/4270-rural-radio-agricultural-extensionthe-example-vernacular-radio-programmes-soil-and-water.

Climate Curriculum. (n.d.). *Climate lessons*. Retrieved from https://www.climatecurriculum.com/climate-lessons-1

Climatelinks. (2017). *Climate risk profile*. Haiti. Retrieved from https://www.climatelinks.org/resources/climate-risk-profile-haiti

Commonwealth Education Hub. (2015). *Education & Climate Change. 29*. Retrieved from https://www.thecommonwealth-educationhub.net/wp-content/uploads/2015/12/Climate-Change-Discussion-Summary.pdf

van der Linden, S., Maibach, E., & Leiserowitz, A. (2015). Improving public engagement with climate change: Five 'best practice' insights from psychological science. *Perspectives on Psychological Science, 10*(6), 758–763. https://doi.org/10.1177/1745691615598516.

Dib, C. Z. (1988). Formal, non-formal, and informal education: Concepts/applicability. *AIP Conference Proceedings, 173*, 300–315.

Evans, G. W. (2019). Projected Behavioral impacts of global climate change. *Annual Review of Psychology, 70*(1), 449–474.

Folke, C. (2006). Resilience: The emergence of a perspective for social-ecological systems analyses. *Global Environmental Change, 16*(3), 253–267.

Friedman, A. (Ed.). (March 12, 2008). *Framework for evaluating impacts of informal science education projects* [On-line]. Available at http://insci.org/resources/Eval_Framework.pdf

Gallagher, K. S., Perry, K., Wansem, M., Kuhl, L., & Frapaise, L. (2019). *Analysis of international funding for Haiti's climate change priorities. SSRN Electronic Journal.* https://doi.org/10.2139/ssrn.3333173.

Hadjadj, B. (2000). *Education for all in Haiti over the last 20 years: Assessment and perspectives.* Kingston: Office of the UNESCO Representative in the Caribbean.

Haiti Partners. (2018, January 18). *Haiti statistics.* Retrieved from https://haitipartners.org/haiti-statistics/

Internews. (2013). *Haiti media and telecoms landscape guide.* Retrieved from https://internews.org/resource/haiti-media-and-telecoms-landscape-guide

Jayaprakash, Y. T., & Shoesmith, B. (2007). Community radio and development: Tribal audiences in South India. In L. K. Fuller (Ed.), *Community media* (pp. 43–53). New York: Palgrave Macmillan.

Khan, M. A. A., Khan, M. M. R., Hassan, M., Ahmed, F., & Haque, S. M. R. (2017). Role of community radio for community development in Bangladesh. *The International Technology Management Review, 6*(3), 94–102.

Læssøe, J., Schnack, K., Breiting, S., & Rolls, S. (Eds.) (2009). *Climate change and sustainable development: the response from education.* A cross-national report from international Alliance of leading education institutes. Retrieved from https://www.dpu.dk/fileadmin/www.dpu.dk/edusudclimatechangeandsustainabledevelopment/documents/om-dpu_institutter_institut-for-didaktik_20091208102732_cross_national-report_dec09.pdf

Ledley, T. S., Rooney-Varga, J., & Niepold, F. (2017). Addressing climate change through education. *Oxford Research Encyclopedia: Environmental Science.* https://doi.org/10.1093/acrefore/9780199389414.013.56

National Education Association (NEA). (n.d.). *Climate change education: Essential information for educators.* Retrieved from: http://www.nea.org//home/65564.htm

Nelson, D., Adger, W., & Brown, K. (2007). Adaptation to environmental change: Contributions of a resilience framework. *Annual Review of Environment and Resources, 32*, 395–419.

Ng Cheong-Lum, R. (2005). *Haiti (Cultures of the World)* (2nd ed.). Tarrytown, NY: Marshall Cavendisk Benchmark.

NOAA. (n.d.) *The essential principles of climate literacy.* Retrieved from https://www.climate.gov/teaching/essential-principles-climate-literacy/essential-principles-climate-literacy

Radical Partners. (2019). *100 great ideas: Climate resilience & sustainability.* Retrieved from https://issuu.com/radical.partners/docs/climate_resilience___sustainability

Roca, C.C. (2016, April 14). Radio in Haiti is the heartbeat of the community. *Miami Herald.* Retrieved from https://www.miamiherald.com/news/nation-world/world/americas/haiti/article71931957.html

Salmi, J. (2000). Equity and quality in private education: The Haitian paradox. *Compare: A Journal of Comparative and International Education, 30*(2), 163–178.

Sewlal, R. (2014). The role and rise of community radio in the development of democracy in South Africa: 20 years of democracy. *Rhodes Journalism Review, 2014*(34), 24–26.

Singh, B., & Cohen, M. (2014). *Climate change resilience: The case of Haiti.* Retrieved from www.oxfamamerica.org/explore/researchpublications/climate-change-resilience-thecase-of-haiti

Slagle T. & Rubenstein, M. (2012, February 1). *Climate change in Haiti.* Retrieved from https://blogs.ei.columbia.edu/2012/02/01/climate-change-in-haiti/

Smucker, G. R., Bannister, M., D'Agnes, H., Gossin, Y., Portnoff, M., Timyan, J., & Toussaint, R. (2007). Environmental vulnerability in Haiti: Findings and recommendations. *USAID/Haiti.*

Stanford Earth. (n.d.). *Climate change education: Curriculum.* Retrieved from https://earth.stanford.edu/climate-change-ed/curriculum

The World Factbook: Haiti. (2018, February 01). Retrieved from https://www.cia.gov/library/publications/the-world-factbook/geos/ha.html

UNESCO. (2010). *The UNESCO climate change initiative: Climate change education for sustainable development.* Retrieved from https://unesdoc.unesco.org/ark:/48223/pf0000190101

UNESCO. (2012). *Climate change education for sustainable development in small island developing states: Report and recommendations.* UNESCO experts meeting 21–23 September 2011, Nassau, Commonwealth of the Bahamas. Retrieved from https://unesdoc.unesco.org/ark:/48223/pf0000216473

UNESCO. (2015). *Not just hot air: Putting climate change education into practice.* Retrieved from https://unesdoc.unesco.org/ark:/48223/pf0000233083

UNESCO. (2019). UNESCO and sustainable development goals.. Retrieved from https://en.unesco.org/sustainabledevelopmentgoals

UNESCO. (n.d.) 'Climate Change Education'. Retrieved from https://en.unesco.org/themes/education-sustainable-development/cce

United Nations Development Program (UNDP). (n.d.). *National adaptation plans in focus: Lessons from Haiti.* Retrieved from https://www.adaptation-undp.org/resources/project-brief-fact-sheet/national-adaptation-plans-focus-lessons-haiti

USAID. (2017). *Climate risk profile.* Haiti. Retrieved from https://www.climatelinks.org/sites/default/files/asset/document/2017_Cadmus_Climate-Risk-Profile_Haiti.pdf

Williamson, K., Satre-Meloy, A., Velasco, K., & Green, K. (2018). *Climate change needs behavior change: Making the case for behavioral solutions to reduce global warming.* Arlington, VA: Rare.

World Population Review. (2020). *Haiti population 2020.* Retrieved from http://worldpopulationreview.com/countries/haiti-population/

Ashley Bazin is a graduate of the Education Policy and Management Masters Program at the Harvard Graduate School of Education. She received her B.A. in Psychology from Rhode Island College and has done research in both developmental and cognitive psychology. Prior to attending HGSE, she worked at a state-wide, youth advocacy organization, for low-income High School students, assisting in their leadership development. Ashley has a vested interest in increasing the access to quality education for both youth and adults.

Christelle Saintis is a graduate of the International Education Policy Masters program at the Harvard Graduate School of Education. She received her B.A. in History from The George Washington University. Prior to her studies at Harvard, she taught Special Education in the Chicago Public Schools before moving abroad and teaching English in Thailand. Most recently, she taught at a bilingual school in Honduras. Christelle is passionate about language education in developing countries and promoting language policies that support bilingual education.

Chapter 5
Adaptation, Migration, Advocacy. A Climate Change Curriculum for Out-of-School Children in Badin, Sindh

Natasha Japanwala

5.1 Introduction

There are two core challenges when it comes to designing a non-formal curriculum for out-of-school youth in a community already vulnerable to the effects of climate change. The first is that the curriculum needs to account for the fact that its intended audience has either never attended school or dropped out after completing only a few years of basic schooling, often learning little during those years. The second is that the curriculum needs to do more than help students create an understanding of climate change—it needs to help them figure out how to survive its effects. In this chapter I present a curriculum designed in partnership with *The Citizens Foundation.* I prepared an environmental sustainability unit as part of their pilot "literacy and life skills" curriculum for out-of-school youth in Badin. These youth are between the ages of 10–14. The program will run for 6 months and is slated to begin in January 2020.

Badin is a case study of a vulnerable community, where the population is facing disproportionate threats from the effects of climate change. According to the United Nations, in 2016 the 15 countries most vulnerable to climate-related disasters were the least developed countries, landlocked developing countries, or small island developing states (Chazanoel and Puscas 2019). These countries are all disproportionately impacted by climate change and are the least equipped to cope because of their lack of resources. The curriculum designed for Badin is low-cost and low-resource, making its framework adaptable to other contexts. It is critical to educate children in vulnerable communities not only about the science behind what they are witnessing, but also enable them to strategize and advocate for their own futures.

Since there are few curricula that focus on climate change education for vulnerable communities, this curriculum will test four ways to react to the effects of

N. Japanwala (✉)
Harvard Graduate School of Education, Cambridge, MA, USA
e-mail: natashajapanwala@gse.harvard.edu

F. M. Reimers (ed.), *Education and Climate Change*, International Explorations in Outdoor and Environmental Education,
https://doi.org/10.1007/978-3-030-57927-2_5

climate change: context, adaptation, migration, and advocacy. The curriculum (see Appendix) dedicates a month each to these four phases, keeping in mind that the pilot consists of other units, that attendance may be spotty, and that teacher capacity may be weaker than anticipated. The curriculum and lesson plans will be designed and translated into the local language, Sindhi.

The theory of action for each phase of the curriculum is simple: (1) if students understand the concept of climate change and can apply it to the phenomena they see around them, then they will be able to articulate what is happening to their community; (2) if students understand what is happening to their community and practice exercises to protect themselves and adapt to the risks posed by natural hazards, then they will be better able to suggest and implement these strategies in their day-to-day lives; (3) if students understand what is happening to their community and have thought about migration and even simulated it, then they will be mentally prepared to move to a new place should their families have to do so; and finally (4) if students understand what is happening to their community and can articulate it, then they will be able to advocate for their rights.

This chapter outlines the four phases of the curriculum, grounding the design of each phase on evidence from both academic research and programs that have been conducted in similar geographical regions. Firstly though, I review the current state of climate change awareness in Pakistan, the particular risks that Badin faces, why educating out-of-school youth is a particular challenge, and the pedagogical methods used to create a curriculum that would meet the needs of out-of-school youth in this particular context.

5.2 A Shifting Attitude Towards Climate Change

Across urban Pakistan from the biggest to the smaller cities, youth are increasingly cognizant of ways in which climate change can wreck their future. It was this awareness that led to the climate march on 20th September 2019, when thousands of students and civilians marched in more than 32 cities, inspired by 16-year-old Greta Thunberg, and demanding action from the current government, the Pakistan Tehreek-e-Insaf (Jamal 2019). Not long after, the journalist Basil Andrews published an essay about rising temperatures in Karachi, the city with the largest population in Pakistan, and how the built environment exacerbates the heat. "I glance at my temperature meter: 42.1 degree Celsius it reads as the black digits stare out at me in static certainty," he wrote. "Climate change is here and whether I, whether we, like it or not, it will get hotter. It already has." (Andrews 2019, page needed).

Since Karachi's 2015 heatwave, which left over 1000 people dead (Imtiaz and Rehman 2015), there has been a rapidly growing acceptance that rising temperatures are wreaking havoc—Pakistan was, according to the Global Climate Risk Index, the 8th country most affected by climate change during 1998–2017 (Eckstein et al. 2018). Different regions face different risks, ranging from temporary and permanent flooding to an increased risk of natural hazards like cyclones and tsunamis, both of which will reduce agricultural production and destabilize an already weak economy, potentially creating millions of climate refugees (Nazar 2016).

Moving from the urban centers to the rural areas, the threats become more severe. First, rural economies, which rely entirely on agriculture, are more susceptible to the effects of climate change. Second, education infrastructure is weak in these areas, leaving the population poorly equipped to devise strategies to mitigate the effects of climate change. This is not to say that the population is unaware of the risks they face, but that they do not have the context to understand how they will need to adapt to those risks. It is for this precise reason that this curriculum focuses on Badin, a rural district in Sindh which is experiencing the effects of climate change right now.

5.3 The Risks Faced by the Population in Badin

Badin lies on the coast of the Arabian Sea in Sindh, to the east of the Indus Delta, and about a 3-h drive from Karachi. It sits at the confluence of almost every kind of climate change risk imaginable: at the height of summer, desert temperatures bring both heatwaves and droughts; since it sits on the coastal belt, it is at risk of being swallowed by rising sea levels, as well as being potentially flattened by cyclones; and finally, increased saline flow from the sea is destroying local ecosystems (including land that was used for agriculture, and breeding grounds for fish). These risks are compounded by the fact that Badin has also suffered a human-induced disaster: between 1984 and 1997, the Left Bank Outfall Drain (LBOD) was built to prevent waterlogging and salinity in three districts in Sindh (Mirpurkhas, Sanghar, and Nawabshah). What ended up happening is that a canal included in the LBOD to carry saline drainage to the Arabian Sea ended up flooding the districts located close to the coast, leaving Badin doubly vulnerable (Rao and Maqbool 2014).

The most severe threat Badin is currently facing is salinization. Researcher Hussain Bux Mallah wrote that it was difficult to gauge not just how many people have migrated as a result of this environmental destruction, but where they went, and whether their migration was voluntary. While he points out that the obvious choice for most people is the nearest big city, Karachi, the reality is that most families are assetless and are forced to either remain where they are or relocate to nearby areas that are geographically similar (Bux Mallah 2012). Given the fundamental impossibility, at this stage and with scant research on the topic, of predicting which course of action families will follow, this curriculum includes both adaptation strategies should they choose to stay and migration strategies should they choose to leave.

5.4 Educating Out-of-School Youth

This curriculum was designed to be sensitive to two factors: firstly, the target demographic, youth in Badin, are actively experiencing the effects of climate change and secondly, that this demographic has never attended school before. The district, even without accounting for the effects of climate change is crippled by poverty: 47.9% of children are underweight and the infant mortality rate is as high as 87 deaths per 1000 births. An indication of how broken education access in the district is: the literacy rate is 39%, the enrolment rate is 41%, the rate of primary school completion is 32%, and as many as 46% of school-age children are out of school (Rao and Maqbool 2014).

Most of the literature currently available on educating out-of-school youth, whether academic or in the form of reports from governments or multilateral agencies, focuses on improving their access to schooling and making the transition to schooling more effective. In order to understand the needs of the target demographic I interviewed Hina Saleem, the Head of Literacy and Life Skills at *The Citizens Foundation*, who is designing and implementing the pilot curriculum for out-of-school youth of which this unit on climate change is a part. "When we met the kids, they weren't even talking to us," she said, while describing a field visit she herself had undertaken to Badin. "We need to work on discipline, forming routines to go to school. Motor skills, as well. How to hold a pencil, how to work with it."

Our conversation illuminated the pedagogical methods that were needed to make this curriculum a success. Project-based learning was a great place to start, since it does not require literacy and builds social skills: students who have never been to school have no literacy skills and struggle to socialize in the collaborative and conversational ways of conventional schooling. In addition, looking at the science of learning more broadly, it was critical to designing a curriculum that was reasonably paced, that would not overwhelm students, and that would present learning to them as something fun instead of tedious.

For example, according to the research-based "gold standard" model for project-based learning from the Buck Institute of Education, effective projects pose a question or challenge ("what it is 'about,' if one were to sum it up – is [it] a problem to investigate and solve, or a question to explore and answer" and that reflection is a critical final piece of an overall project ("reflection can occur informally, as part of classroom culture and dialogue, but should also be an explicit part of project journals, scheduled formative assessment, discussions at project checkpoints, and public presentations of student work" (Larmer 2015). During the first phase of the curriculum, students complete a simple science experiment to understand the three major risks their community faces—these are framed as questions instead of as topics ("Why is it hot?" "Why does the sea flood?" and "Why is salt bad for plants?"). "Circle time" is built into the end of each lesson plan, to facilitate time for reflection, but also to include more space for students to develop their oral language and social interactions skills.

The curriculum was designed with attention to some of the findings of the science of how people learn. For example, one of the core cognitive principles governing learning is that students best learn new concepts by referring to ideas they already know—for this reason, the materials proposed to help students understand climate change in their particular context are based on what they can physically see around them in their community. The curriculum also moves slowly and allows time to revise concepts, since another core cognitive principle is that students' understanding of new ideas can be inhibited if they are confronted with too much information at once (Deans for Impact 2016).

A review of the existing literature on climate change education, helped develop a narrative arc for the curriculum. A review of climate change education strategies by Martha Monroe et al. (2017) makes a helpful distinction between teaching "the facts" of climate science and "actions" to mitigate climate change—the authors distinguish between educators who "believe their job is limited to conveying factual information about climate science" and others who "intend to build critical thinking skills and help youth understand the sources of conflict about climate change or prioritize problem solving skills as they help youth conduct local projects to mitigate and adapt to climate change" (pp. 791–812).

The four phases of the curriculum are designed to move students from understanding to action. While the first phase focuses on helping students understand the facts of climate change, the remaining three focus on project-based activities that help them live with and survive the impact climate change is having on their communities. Studies have suggested that project-based learning fosters higher engagement and academic gains, even in disadvantaged populations (albeit in a context vastly different from Badin) (Halvorsen et al. 2012). Drawing inspiration from approaches designed by non-profits like Aflatoun, which run entrepreneurship programs for children in rural Rwanda (Aflatoun International 2019), or Dream a Dream, which works with out-of-school children in India (Dream a Dream 2020), this project-based curriculum encourages students to draw on their community for research and understanding, and work collaboratively to develop adaptation and survival strategies for their own futures is the best way forward.

Phase 1: Context

The first step of the curriculum is helping students understand what climate change actually is, rather than instructing them about climate change as it is occurring around the globe, the introductory unit of the curriculum focuses on three phenomenon that impact Badin directly and that the students will have observed around them: overheating in the summer, floods because of sea levels rising, and the lack of productivity of the land because of salinization. This is largely in line with the overarching philosophy driving climate change curricula: for example, the US-based National Center for Science Education (2016) describes four best practices for climate change education that can be considered universal: "Make it local, make it human, make it pervasive, make it hopeful."

In order to communicate the three phenomena to students, the curriculum relies on two strategies: a pictorial flip-book, and a simple experiment.

An innovative way to explain the impact of the greenhouse effect, flooding, and salinization to students in Badin is to use a pictorial flip-book—an idea inspired by Save the Children's "community strategies flip-book", which was an intervention to encourage community efforts towards literacy (Bowd et al. 2010). Each page contained an action parents and children could complete to support the development of reading skills. While our flip-book will not have actions, it will have a brief sequence of images through which students can piece together the story of climate change with guidance from their teachers. The teachers will, in turn, be supported via their mobile phones—an increasingly popular teacher professional development intervention in low-resource settings (Burns 2013). The combination of the picture flip-book and live audio support for teachers will allow the community to collectively understand and articulate what they are seeing play out around them. The physicality of the flip-books would also allow students to take the stories home and share them with their parents or siblings who aren't enrolled in the life skills program.

For example, for the lesson on the flooding, the first image will show a mountain with a bright sun over it, the second will show a river flowing down the mountain towards the sea at its foot, and the third will show the sea overflowing and inundating the land. These images will be bold and will not be accompanied by any text, but will give students a chance to discuss among themselves what the images mean. This bolsters their communication skills, while also allowing them to grapple with concepts that will be cemented through a follow-up experiment. The experiment will physically demonstrate the concept they explored in the flipbooks—in this case, it will demonstrate how a block of ice melting raises the level of water in a bucket.

Phase 2: Adaptation

To understand how the citizens of Badin should cope with the effects of climate change, it is helpful to look at the literature produced by BRAC, the Bangladesh-based international development organisation thought to be one of the largest in the world. The risks faced by Bangladesh are similar to those faced by Badin. Two-thirds of the country is less than five meters above sea level, which means that saltwater intrusion from sea level rise renders, as it does in Badin, agricultural workers unemployed while also creating a situation of food insecurity (BRAC Climate Change Programme, 2017).

BRAC's mission in Bangladesh is to prepare people to adapt and respond to the effects of climate change, while also mitigating future impacts through sustainable development practices (BRAC Climate Change Programme 2017). There's a distinction between adaptation and mitigation—adaptation is helping people to cope with the effects of climate change, while mitigation is equipping them to prevent further effects. In the case of Badin, where the effects are already being felt, it makes sense to focus on adaptation. That said, it's important to note that not all adaptation strategies can be taught to 10–14 year olds. For example, one adaptation strategy in Bangladesh has been to invent salt-tolerant rice varieties for coastal areas where soil salinity is high. Other adaptation strategies are easier for 10–14 year olds to adopt, like constructing housing with raised plinth levels and using elevated latrines, both to cope with flooding. Both of these strategies provide opportunities to engage in tactile and collaborative project-based learning—teachers could give students rudimentary construction materials, like paper and playdough, and present them with the challenge of building a house to withstand flooding. They could do the same exercise with the elevated latrines, this time using the lesson to explain the importance of limiting the spread of diseases.

In an example lesson, students engage in the design thinking process to construct levees along a river bank. The challenge is that they are given different materials and they have to decide on the materials and design that will lead to the strongest levee. The levee exercise builds skills that will be directly useful—a case study by Oxfam International, which summarizes the findings from community-based research in Badin in 2009, recommended strengthening the lining of existing water ponds, constructing embankments to protect fields from seawater intrusion using pressed earth, and constructing emergency flood shelters (Oxfam International 2009).

Using design thinking and project-based learning to teach adaptation strategies has value beyond familiarizing the students with new skills and the ability to work collaboratively towards a common goal. It builds their leadership skills by encouraging them to think of themselves as climate change ambassadors in their community, especially since the exercises follow a unit where their conceptual knowledge was built. They have, through units 1 and 2, had both the experience of leadership and the vocabulary to articulate the cause they are leading. A publication by the United Nations Joint Framework Initiative on Children, Youth, and Climate Change (2013), titled *Youth Action on Climate Change: Inspirations from Around the World,* contains more case studies about adaptation and disaster risk reduction programs for vulnerable communities, that have been designed specifically to provide climate leadership opportunities for youth. The report makes the case that youth involvement in climate change adaptation is critical for the entire community's survival.

There is potential that once the program wraps up, the students will use what they have learned to be active leaders in their community—a productive follow-up program would be one where they partner with elders in their community or with development organizations or non-profit organizations working in the area on a real project. There are examples of programs where youth in vulnerable communities are given leadership roles on projects. The *United Nations Joint Framework Initiative on Children, Youth, and Climate Change* (2013) publication contains a case study detailing a collaboration between the United Nations Environment Programme and the Municipal Council of Xai-Xai City in Mozambique in January 2010. Here, the goal was to help the community adapt to topsoil erosion resulting from higher rainfall, and one project was building barrier walls that held soil in which resilient plant species were placed. Another case study, this time in collaboration with the El Mouddaa community in Morocco, involved youth building rock dams to stem the flow of floodwater. These are both inspirational projects for what could happen in Badin once this pilot curriculum concludes.

Phase 3: Migration

There is a real possibility that the citizens of Badin will be forced, at some point, to migrate elsewhere. Preparing students for migration can take one of two forms: the first is to prepare them emotionally, and the second is to prepare them practically.

To prepare emotionally for migration, students can engage in simple activities that give them a chance to preserve memories and stories through community interviews and a time capsule. Migration causes various levels of mental distress, beyond just physical displacement—as Dinesh Bhugra (2004) writes, migration includes "experiencing at first a sense of loss, dislocation, alienation and isolation, which will lead to processes of acculturation." (p. 135). In order to help children cope with these potential feelings of loss, what if one lesson of the curriculum was the creation of a time capsule to help them remember who they are and where they came from? This could involve filling a shoebox with meaningful objects, photographs of their home taken with a disposable camera, and maybe even drawings about what and who they love most.

On a more practical note, moving to a new place—either a neighboring rural district or to a big urban center—requires both confidence and resourcefulness. One way to prepare students for navigating a new home, especially when their social skills are limited, is to play a scavenger hunt-style game where they work individually to navigate an imagined landscape through a pictorial map, learn how to convey information reliably, make decisions, and interact with new people.

Phase 4: Advocacy

For the final phase, it is important to teach students how to advocate for themselves—not just so that they can demand better resources and negotiate if they have to migrate, but because they should be able to conduct public protests or speak to local authorities if need be. One of the reasons why Badin has suffered as much as it has is because of the human-induced disaster of the Left Bank Outfall Drain.. During this phase, students will learn how to negotiate, how to speak publicly in front of a crowd, and how to move an audience through basic theatre.

Theatre has actually proven to be a successful pedagogical strategy in rural communities in Pakistan. The self-funded and Karachi-based Pakistan Fisherfolk Forum (PFF) use theatre as an advocacy tool in rural communities, drawing on traditional songs and folklore to raise awareness about issues such as shortages of water for agriculture, frequent floods and droughts (reference needed). In an interview with Reuters in 2015, Shafqat Aziz, a food security expert with Oxfam Novib, said the plays were empowering and motivated communities to ask government representatives for policies that addressed food security. "The awareness drives affected communities to a decision-making position and this is where they try to come up with effective solutions for tackling climate change," he said (Saeed 2015).

Given its popularity in rural communities, it will make a fun final lesson to have a community theatre challenge for the students—this will also double as a pleasant way to test them on the concepts they learned through their pictorial flip-books and science experiments in Phase 1 of the curriculum. Students will also have to work in bigger groups than in their previous exercises, which means that they will have to work harder to lead and collaborate. The final unit is both a way to assess whether they have retained concepts taught earlier, but also whether they are able to collaborate and communicate, as covered in Phases 2 and 3 of the curriculum.

5.5 Conclusion

Through drawing on creative project ideas like pictorial flip-books, constructing models of latrines and rock dams, and creating time capsules, it is possible to translate competencies that are serious in their implications (disaster management, migration strategies) in ways that make for a learning experience that encourages both curiosity and enjoyment. It is impossible to think about climate change without also thinking of virtually every other variable in a community: climate change is inherently tied to shelter and sustenance, and it disproportionately impacts the differently abled, children, women, and the elderly. A curriculum for out-of-school youth in a community already vulnerable to climate change is an opportunity not only to develop adaptation and migration strategies, but also to subtly start enabling a new generation to own their future.

Appendix: The Curriculum

This four-phase curriculum has been designed to last 4 months, with a month dedicated to each phase. This is assuming that only one lesson is taught per week. Each phase can also last a week, if lessons are taught each day—this would make it a month-long curriculum. The timeline is deliberately flexible, to allow it to be adapted to different programs and unexpected changes in schedule. Below is an overview of the curriculum, outlining the objective and strategy of each of the four phases. A sample of a lesson plan for each phase follows.

Phase 1: Context

Objective: To understand the concept of climate change in relation to the three biggest risks to the local community: overheating, flooding, and salinization

Strategy: Use flip-books to develop a narrative around what is happening and experiments to understand the science behind events

Phase 2: Adaptation

Objective: Engage in hands-on projects to understand how to adapt to changes in the environment—these will also familiarize students with the principles of design thinking and build social skills via group work

Strategy: Create small-scale projects for students to learn how to put strategies to adapt to the effects of climate change into action

Phase 3: Migration

Objective: Think about the emotional cost and resilience of climate change migration through community interviews and time capsules; build self-resiliency for a life of change; prepare to adapt to new situations with ingenuity and flexibility

Strategy: Lean on the community to build a "bank of memories", draw maps, and create a time capsule (these will also build interaction skills and drawing skills); design a game to simulate arriving in and navigating a new place

Phase 4: Advocacy

Objective: Amplify human stories in an effort to learn how to articulate issues faced to non-community members. This will also develop public-speaking skills and build general confidence

Strategy: Use speeches and theatre to get students to both revise concepts learned in earlier phases and to articulate their problems with clarity and confidence

Phase One: Understanding Climate Change
Title of Lesson: Why does the sea flood?

Idea: What causes the sea to overflow and flood the village?

Objectives (Understanding): To understand how the effect of heating from Lesson 1 ("Why is it hot?") causes ice in the ocean and on the mountains to melt. To understand why melting ice causes sea levels to rise. To understand why a rise in sea levels causes a flood.

Objectives (Skills): Discipline and patience while the teacher demonstrates the experiment.

Important Vocabulary: Heat | Ice | Melt | Flood

Materials Needed: A shallow glass dish or bowl | Cold water | Ice cubes | A marker; Pencil; Paper | A makeshift stove to heat the water

Lesson Structure

Introduction	Explain that ice can be found in two places: On the mountain tops and in the sea. When it gets hot, this ice melts. The melting ice causes floods. Connect these concepts to the pictorial flip-book by asking the students to look at the images and revise the discussion they had during the previous lesson. Explain that you will now demonstrate to them how the ice melts and causes floods.
Activity	1. Take the shallow dish and fill it halfway with water. 2. Put the ice cubes in the dish. Immediately after, mark the level of the water. Explain that this is the regular sea level. 3. Allow the ice cubes to melt naturally. Mark the new "sea level". 4. Now repeat the experiment by filling the shallow dish halfway with water, putting in ice cubes, and marking the level of the water. 5. This time, instead of letting the ice melt naturally, heat it by holding the dish above the makeshift stove. Mark the new "sea level". 6. Compare the marks of the post-melting "sea level" when the ice melted naturally compared to when the ice was heated.
Reflection	Ask students what was different between both the experiments. Keep prompting them without providing the answer. Once they identify that the second time there was heating involved, ask them whether more heat from the sun will cause more flooding. Wait for them to say yes! They should come to this conclusion on their own.
Assessment	Once the experiment is over, bring out the pictorial flip-books once more. Have the students look at the pictures. See if their understanding has become more clear after the experiment. (This should also be a measure of the experiment's success as a teaching strategy.)

Phase Two: Adaptation
Title of Lesson: How can we protect ourselves from floods?

References: Sciencing "A Flood Experiment to Do for Children" https://sciencing.com/flood-experiment-children-10025793.html

---

Idea: What can we do before a flood comes to protect ourselves from its effects?

Objectives (Understanding): To understand that construction efforts, particularly using certain materials, can limit the damage caused by flooding.

Objectives (Skills): Engage in the design thinking process in a very elementary way (asking questions, designing, testing, and iterating); learn to work collaboratively in groups.

Important Vocabulary: Flood | Levee | Wall | Material | Construction | Protect | Safe

---

Materials Needed: A shallow box | Modeling clay | Popsicle sticks | Card paper | Small jug | Water | Small toy houses, made of either wood or plastic

Instructions: Set up the exercise by building a village inside each box: use the modeling clay to create a surface covering the base of the box and then dig a "river" in the middle. Fill this river with water.

Lesson Structure

Introduction	Unlike the previous lesson where explanations were offered, with this lesson do not provide any background or information. Let curiosity get the better of the students. Divide the students into groups of 3–5. Give each group one of the "villages" complete with a river with water in it.
Activity	1. Give each group a set of toy houses and ask them to stick them in by the "banks" of the river. 2. Now tell them a big flood is going to come and show them the jug filled with water. Before the flood comes though, they have to build a wall to protect the houses from the water. 3. Hand out popsicle sticks, card paper, and more modeling clay, but provide no further instructions. 4. Let the students build "walls" to protect their "houses" from the "flood". 5. Once they are done, go around and "flood" each "river" with water from the jug. Have each group watch and judge which "walls" are the strongest.

Reflection	Ask students which "walls" they thought were the strongest. Ask them why. Ask them whether they think the material used makes a difference. Then move into a social-emotional discussion. Ask them whether their homes have been destroyed in a flood previously. Ask them whether their families did anything differently after the incident. Ask them whether they are scared of floods. Ask them whether the experiment changed the way they think about protecting against floods. Is there anything they will do differently from now on? Will they share what they learned with their families?
Assessment	Once the experiment is over, ask students what else they can do to prepare for floods. The design thinking process will hopefully have inspired them to think outside the box. Note who, if anyone, comes up with an idea different from what was discussed and shared (walls and levees).

Phase Three: Migration
Title of Lesson: The "New Place" Game

Idea: Building navigation and social skills so that students can help their families settle into a new home in a new place, should they be forced to migrate.

Objectives (Understanding): To understand that moving to a new place requires being resourceful: this means being aware of one's surroundings but it also means actively seeking new information.

Objectives (Skills): Learn how to navigate by reading a pictorial map; learn how to relay information reliably; learn how to make decisions when confronted with options; and learn how to interact with strangers.

Set-up A map with three icons placed in a rectangle (one of a medicine bottle, another of a bus, and another of a person). As shown in the map, students need to go to each of the three stations as demonstrated by a set of arrows. At the first station, they are told a set of symptoms and they have to ask the "pharmacist" what medicine best fits those symptoms. They then have to decide whether or not to follow the pharmacist's advice. At the second station, they are given a destination and they have to ask the "station master" how to get to that destination. The Stationmaster will give them a set of three different options with three different prices and route times—they then have to decide which route is the most economical. Finally, they have to go to the third station, where they have to "meet" their new neighbour who will ask them personal questions about themselves (age, where they have come from, what they want to be when they grow up…).

Reflection This is not intended to be as intense a lesson as others. The point of this exercise is to get students to think about the challenges they may face in a new place and how to navigate those responsibly, by thinking through the information they receive and communicating clearly and effectively. It also builds core skills like reading a map and relaying information reliably. As a reflection at the end of the exercise, ask the students which of the three stations they found most challenging or intimidating, and what they would do differently if they had to repeat the exercise.

Phase Four: Advocacy
Title of Lesson: Community Theatre Challenge

Idea: To test whether students have retained the concepts taught in Phase 1 and whether their collaboration and communication skills have improved as a result of the activities covered in Phase 2 and Phase 3.

Objectives (Understanding): To demonstrate a basic understanding of one of the three concepts covered in Phase 1.

Objectives (Skills): To collaborate with others in terms of assigning roles and directing an elementary dialogue-reliant production, and to communicate effectively in delivering ideas.

Set-up A makeshift stage demonstrated only by a rug or even a rectangle of land chalked out with a border.

Lesson Structure

Introduction	Divide students into three groups—and given each one the following prompt: "Why is it hot?" "Why does the sea flood?" and "Why is salt bad for plants?" Tell them that they have to come up with a brief skit or play around their topic. Explain that they can be creative and need not repeat what they remember from their pictorial books or the experiment directly.
Activity	1. Let students spend half the lesson preparing their skit. 2. During the other half, stage all three skits.
Reflection	Use this opportunity to refresh concepts if one of the groups conveyed information that could have confused the others. Congratulate the students on their performances, and ask them how they felt about it and whether they enjoyed it.
Assessment	Take notes while students are both preparing and performing as to who is speaking, who is leading, and who is doing. See if their confidence and communication has improved.

References

Aflatoun International. (2019). *Hatching a plan in Rwanda.* Aflatoun International. Retrieved from https://www.aflatoun.org/stories/hatching-plan-rwanda/

Andrews, B. (2019, October 3). *Fahrenheit Karachi.* Dawn. https://www.dawn.com/news/1508567/fahrenheit-karachi

Bowd, A. J. et al. (2010). Community strategies for promoting literacy. Save the Children.

BRAC *"Climate Change Programme."* BRAC. http://www.brac.net/program/climate-change-programme/

Burns, M. (2013, August 20). *If we 'text' it, will they learn?* Global Partnership for Education. https://www.globalpartnership.org/blog/if-we-'text'-it-will-they-learn

Chazanoel, M. T., & Puscas, S. (2019). *Climate change and migration in vulnerable countries.* United Nations. www.un.org/sustainabledevelopment/blog/2019/09/climate-change-and-migration-in-vulnerable-countries/

Deans for Impact. (2016, December). *The science of learning.* Deans for Impact. Https://Deansforimpact.org/Wp-Content/Uploads/2016/12/The_Science_of_Learning.Pdf

Dream a Dream. (2020). *Our strategy.* Retrieved from http://dreamadream.org/our-strategy

Eckstein, D. et al. (2018). *Global climate risk index 2019.* Germanwatch.

Halvorsen, A., et al. (2012, August 3). Narrowing the achievement gap in second-grade social studies and content area literacy: The promise of a project-based approach. *Theory & Research in Social Education, 40*(3), 198–229.

Imtiaz, S., & Ur-Rehman, Z. (2015, June 25). Death toll from heat wave in Karachi, Pakistan, hits 1,000. *The New York Times.* https://www.nytimes.com/2015/06/26/world/asia/karachi-pakistan-heat-wave-deaths.html

Interact. Arctic climate change "What causes the sea level to rise?". INTERACT and Wicked Weather Watch. https://eu-interact.org/app/uploads/2017/10/Demonstrating-Arctic-Climate-Change_Rising-sea-levels.pdf

Jamal, S. (2019, September 9). Pakistani students, youth join global climate change campaign. *Gulf News.* https://gulfnews.com/world/asia/pakistan/pakistani-students-youth-join-global-climate-change-campaign-1.66554299

Larmer, J. (2015, April 21). *Gold standard PBL: Essential project design elements.* Buck Institute for Education: PBL Works. https://www.pblworks.org/blog/gold-standard-pbl-essential-project-design-elements

Mallah, H. B. (2012, December). *Social inequality and environmental threats in Indus Delta Villages: Pakistan.* Paper presented at the ESF-UniBi-ZiF conference on "tracing social inequalities in environmentally-induced migration" Center for Interdisciplinary Research, At, Bielefeld, Germany.

Monroe, M. C. et al. (2017, August 13). Identifying effective climate change education strategies: A systematic review of the research. *Environmental Education Research, 25*(6) 791–812.

National Center for Science Education. (2016). *Teaching climate change: Best practices.* Retrieved from https://ncse.ngo/teaching-climate-change-best-practices

Nazar, S. (2016, March 4). *Pakistan's big threat isn't terrorism—It's climate change.* Foreign Policy. https://foreignpolicy.com/2016/03/04/pakistans-big-threat-isnt-terrorism-its-climate-change/

Oxfam International. (2009, November). *Climate change adaptation in practice: Community-based adaptation and advocacy in coastal Pakistan.* Oxfam International. https://www.oxfam.org/sites/www.oxfam.org/files/pakistan-climate-adaptation.pdf

Rao, Z., & Maqbool, I. (2014). *Climate change risks and vulnerabilities of Badin District in Sindh Province, Pakistan.* ADPC and Oxfam Novib.

Saeed, A. (2015, November 26). As climate impacts hit, Pakistan faces migration surge: Experts. *World News.* https://www.reuters.com/article/us-pakistan-climatechange-migration/as-climate-impacts-hit-pakistan-faces-migration-surge-experts-idUSKBN0TG02J20151127

Natasha Japanwala is a writer and teacher from Karachi, Pakistan. She has worked with K-12 schools developing and implementing curricula focused on storytelling, design thinking, and local problem-solving for primary, middle, and secondary students. In 2017, she founded The Infinity Project, a non-profit committed to teaching writing to youth and women from vulnerable communities. As a freelance journalist, she writes widely about Pakistan's burgeoning youth demographic for publications across the world. Her reportage and commentary has appeared in The Washington Post, The Independent, and The Telegraph, among others. She is a graduate of the International Education Policy Masters program at the Harvard Graduate School of Education and a B.A. in English Literature and Creative Writing from Princeton University.

Chapter 6
Students as Partners. Implementation of Climate Change Education Within the Harvard Graduate School of Education

Annie Hyokyong Nam and Sueyoon Lee

6.1 Beyond the Bottom-Up and Top-Down Debate on Climate Change Education

Climate is a dynamic interplay of the atmosphere, hydrosphere, cryosphere, biosphere, and lithosphere (Aspen Global Change Institute 2019). Whereas weather is defined by fickle fluctuations from day to day, climate change is a long term, sustained trend of change in climate. As rising anthropogenic greenhouse gas emissions continue to trap heat in the form of infrared radiation (Fahey 2012), global warming has been occurring at rates much faster than anticipated and its effects being clearly felt worldwide.

Despite alarming amount of evidence from the scientific community, global warming remains an issue of political debate in the United States, accentuating the ideological phenomenon categorized as 'climate change denial (Hess and Collins 2018).' The withdrawal of the United States from the 2015 Paris Agreement Treaties (Pompeo 2019) and the reversal of numerous environmental rules and regulations (Clements et al. 2020) indicate that American politics may stand in the way of achieving an environmentally sustainable future. Consequently, local and regional actors in the US are increasingly stepping forward to fill the policy void created by federal inaction (Reeves et al. 2014). Inciting and mobilizing citizen action could be central in mitigating the effects of climate change (Wi 2019).

If students are educated to understand and act upon changes in climate, they can create a grassroots movement that produces systemic changes. A grassroots movement refers to an initiative to help individuals engage in community interventions and activities with the mission of instrumenting local and societal change for the collective interest of the community (de Souza 2007; Fisher 1998; Rothman 1996).

A. H. Nam · S. Lee (✉)
Harvard Graduate School of Education (HGSE), Cambridge, MA, USA

F. M. Reimers (ed.), *Education and Climate Change*, International Explorations in Outdoor and Environmental Education,
https://doi.org/10.1007/978-3-030-57927-2_6

Grassroots organizations have shown promising results in raising awareness and encouraging community involvement (Christens 2010; People's Association [PA] 2011; Paul and Tan 2003; Smith 2000), but the debate over top-down versus bottom-up approaches in climate change education (CCE) has long persisted (Fadeeva et al. 2014).

In essence, the main aim of climate change education (CCE), which is rooted in education for sustainable development (ESD) (UNESCO 2015), is to engage different stakeholders in promoting lifelong education for global citizenship and help build a knowledge society in which local communities act upon recommendations. Educating for this type of change is a challenge in traditional western education because education is no longer considered a top-down transmission of knowledge, information, and values (Varga et al. 2007). Rather, a bottom-up approach promotes creativity and ownership of joint involvement and action, allowing for a leveraging of specialized knowledge and experiences contributed by citizens (Fraser et al. 2006; Wi 2019). However, there are some advantages to a top-down educational approach. According to Wi (2019), its simplicity and efficiency in decision making as well as its ability to mobilize more resources and generate widespread awareness can help get things done. Ultimately, Wi argues for a collaboration and infusion of both approaches as key to a successful implementation of CCE.

Literature additionally points out the potentially fruitful and productive partnerships of grassroots initiatives with large scale institutions and/or government organizations. Reeves et al. (2014) suggests governments should set up grassroots initiatives themselves 'from the outside' to generate climate change activism among citizens. Within higher education institutions, Fadeeva et al. (2014) talks about the need to move beyond the top-down versus bottom-up participation and embrace a participatory democracy in which everyone is involved and held responsible. Brundiers et al. (2014, p. 196) comment that to move beyond the inertia of academic institutions and break established patterns requires an innovative, multilateral relationships between faculty, students, and surrounding communities.

In this chapter, we explore the question acutely posed by Van der Leeuw et al.

> Academic institutions remain so inertial because the professoriate remains in familiar and comfortable patterns. This is human nature but denudes the academy of the energy and passion needed for change. A more bilateral relationship between faculty and students might produce different outcomes. *If students played an equal role in the development of curricula, selection of course content, and initiation of applied projects, how different might the impact of the academy become?* (Van der Leeuw et al. 2012, p. 118).

We begin by examining the necessary design elements of embedding CCE within schools of education and then articulate the implementation process of a CCE curriculum at the Harvard Graduate School of Education (HGSE). Last, we discuss the limitations and implications of "students as partners" in the co-creation of learning and teaching in higher education institutions and in the field of sustainable development education.

6.2 What to Consider When Integrating Climate Change Education (CCE) Within Schools of Education

Previous efforts to embed CCE within education systems have been centered around teacher education (Vega-Marcote and Varela-Losada 2016; Hopkins and McKeown 2014; Varga et al. 2007) or a school-based implementation of a whole school approach (UNESCO 2014). However, neither teachers nor schools are the only constituents of education. Multiple key stakeholders in education such as educational policy makers, researchers, teacher certification boards, teacher education institution executives, administrative staff, and students must all work together in the reorientation towards sustainability. Ferreira et al. (2007) call this a "Whole-of System Model," in which all elements of the system are engaged in the process of embedding a new curriculum, new policy, or whatever is (collectively) determined to be most needed within the particular contexts.

The difficulty of serving the whole of the education system (Vega-Marcote and Varela-Losada 2016) has long been cited as an existing tension plaguing the current environmental and sustainability education efforts in pre-service and in-service teacher education. Graduate schools of education are hubs that attract pre-professionals and professionals from various roles in education. A CCE curriculum that facilitates the collaboration of pre-professionals and professionals across the education sector would promote peer-learning and collaboration, infusing the diversity of perspectives and modeling a whole-of systems approach.

To note, the desire for a curriculum like ours arose not because we wanted to discourage or diminish the existence of CCE curriculum already blended implicitly within several existing structures in schools of education (Denby and Rickards 2016; Molthan-Hill et al. 2019), but to draw out an explicit curricular experience of CCE and model a new multi-lateral approach among students at schools of education. Too often, when students engage in classes in which elements of sustainable development (SD) are implicit within the content or pedagogy of the class (Denby and Rickards 2016), key competencies in sustainable development (SD) are passed down in an 'unconscious' or 'unofficial' way, making the transition from knowledge into action difficult (Lambrechts et al. 2013).

What emerged was the idea to create a curriculum with the blend of explicit content knowledge and purposeful pedagogical structure that was partially student led but faculty supported. A successful implementation of a curriculum within one school of education could generate momentum for other schools of education to follow suit.

In order to follow a comprehensive approach to integrate much of what is known about how educational institutions change to become more relevant, we examined the conception of a student led curriculum through a framework that analyzes the process of educational change to advance global education through five perspectives: cultural, psychological, professional, institutional, and political (Reimers 2020).

6.2.1 A Cultural Perspective of a Student Led Curriculum of Climate Change

A cultural perspective emphasizes that educational practice is the result of shared norms, artifacts and practices which define how education is broadly understood in a society (Reimers 2020). Researchers have found that most people who recycle do not do so out of concern for the environment; they do it because it is socially encouraged – a sociocultural norm (Schultz 2002).

Climate change involves global problems that require social, technological, and political relations to be successful (Räthzel and Uzzell 2009; Wi 2019). CCE also hinges upon a shared vision defined by innovative, multilateral relationships among faculty, students, and surrounding communities (Brundiers et al. 2014). Establishing a shared sense of environmental responsibility is a critical cultural norm, serving as the foundation for numerous sustainability initiatives.

Another way to introduce an inclusive culture of sustainability is to land on a shared meaning of terminology. CCE is inundated with terminologies that are similar in nuance but different in meaning (Varga et al. 2007) which can be intimidating and confusing. Moreover, global standards and guidelines on a common CCE have yet to be implemented.

6.2.2 A Psychological Perspective of a Student Led Curriculum of Climate Change

Next, a psychological perspective highlights the implications of knowledge about how people learn for the process of changing teaching and learning for students, teachers and others supporting instruction (Reimers 2020). This means that education programs need to be purposefully designed, with specific behavioral changes targeted from the outset. It is not enough for students to acquire theoretical concepts but also to be prepared to act, based on sensible decisions for real-world and complex situations (Vega-Marcote and Varela-Losada 2016). This implies that the educational actions should seek the development of specific skills that foster sustainable actions. Wiek et al. (2011) illustrates five basic competencies of education for sustainable development (ESD) that should be combined to reach this aim:

- **Systems-thinking competence:** the ability to collectively analyze complex systems across different domains (society, environment, economy, etc.) and across different scales (local to global), thereby considering cascading effects, inertia, feedback loops and other systemic features related to sustainability issues and problem-solving frameworks.
- **Anticipatory competence:** the ability to collectively analyze, evaluate, and craft rich visions of the future related to sustainability issues and problem-solving frameworks.

- **Normative competence**: the ability to collectively map, specify, apply, reconcile, and negotiate sustainability values, principles, goals, and targets.
- **Strategic competence:** the ability to collectively design and implement interventions, transitions, and transformative governance strategies toward sustainability.
- **Interpersonal competence:** the ability to motivate, enable, and facilitate collaborative participatory sustainability research and problem solving.

Moreover, best practices in education and effective pedagogies must be infused throughout the design of a competency-driven curriculum. We provide a prototype of a competency driven curriculum embedded with the best practices of twenty-first century learning in section 6.3 (stage 5) and in the Appendix B.

Finally, climate apathy is real and students themselves often do not feel the need to learn or take a course on climate change. Psychologists Kasser and Ryan (1996) define four types of motivations propelling individuals towards a goal. Extrinsic motivation as well as intrinsic ones can incite motivation among students. To incite intrinsic motivation and a connection to nature, Hungerford and Volk (1990) introduce the importance of fostering a personal connection to nature and taking ownership of a problem. Extrinsic motivation can be cultivated when elements of CCE are seen to translate into very practical elements in the workforce. In 2017, the sustainability sector saw a substantial increase in employment as well as average wages $5000 above the national median (Environmental Defense Fund and Meister Consultants Group 2017).

6.2.3 A Professional Perspective of a Student Led Curriculum of Climate Change

The professional perspective goes beyond teacher pedagogy and focuses on the structure of roles and institutions in integrating expert knowledge into practice (Reimers 2020). Teachers are an essential pillar in CCE, for they are directly responsible for the teaching and learning process. There is plenty of literature on the impact of teacher-student relationships in the classroom (Forbes and Zint 2010; Roorda et al. 2011; García Bacete et al. 2014), as well as the importance of teachers as role models for the development of environmental literacy (Rickinson 2001; Stern et al. 2010).

Beyond providing relational teaching, instructors play a pivotal role as knowledge building catalysts. To aid in these efforts, instructors can carefully consider the selection of resources made available at higher education institutions that promote professional development initiatives. Instructors often have access to an abundance of institutional resources related to climate change, as universities are often the hub of innovative research and cutting-edge technology (Dyer and Andrews 2011).

Furthermore, there has been rising interest in research and practice about 'students as partners' through co-created learning and teaching (Cook-Sather et al. 2014; Dunne 2016; Mercer-Mapstone et al. 2017). This bilateral relationship often results in greater learning outcomes and skill enhancements for students.

6.2.4 An Institutional Perspective of a Student Led Curriculum of Climate Change

An institutional perspective focuses on the educational structures, norms, regulations, and organizational design (Reimers 2020), exploring the work of teaching and learning through a systemic lens. Numerous challenges and internal/external pressures exist for schools of education to bring forth change on a systemic level.

While CCE is important to all education levels, from primary schools to universities (Harker-Schuch 2019), higher education marks itself distinct in its role in deepening knowledge (Radaković et al. 2017; Vettori and Rammel 2014), inspiring scientifically rigorous expertise (Anderson 2012), and teaching the skills of integration, synthesis, and systems-thinking to cope with complex problems in confronting sustainability challenges (Stephens et al. 2008).

As institutions, schools of education should understand their role in translating cutting-edge research and best practices in education into information that policymakers can apply (Dyer and Andrews 2011). Furthermore, institutions should provide support to grassroots initiatives in SD, for without its support, innovations are short-lived and unable to last beyond personnel and changes within (Hopkins and McKeown 2014).

6.2.5 A Political Perspective of a Student Led Curriculum of Climate Change

A political perspective recognizes that education affects the various interests of many different groups that are often in conflict with one another (Reimers 2020). The role of universities is especially important in areas of the world where strongly opinionated leaders and media outlets have created confusion among the public about environmental issues like climate change. The topic of climate change resonates with deeply held values, such that adults respond by protecting their group identity and ways of life (Monroe et al. 2017).

In the United States, there is a considerable difference between the probability of taking at least one climate-change related course at public research universities with Democrat-controlled state legislatures versus Republican or split-controlled state legislatures (Hess and Collins 2018). The effects of politics trickle down into teacher and student readiness and belief in teaching and learning about climate change. Stevenson et al. (2016) note that a teacher's belief that global warming is happening appears to predict a student's belief.

Despite political challenges, higher education institutions can and must continue to provide an important source of countervailing institutional power to misinformation and lack of policy support when it comes to climate change.

6.3 A Case Study: Implementations of a Student Led Curriculum at the Harvard Graduate School of Education (HGSE)

There were three main reasons to use Harvard Graduate School of Education (HGSE) as a pilot school to test the concept of a student led curriculum.

First, at the time of writing we had been graduate students at HGSE, allowing us easier access to the resources, staff, and administration needed to design a student driven curriculum within a higher-ed institution.

Second, the intent of our curriculum and our efforts to scale aligned with the sustainability mission of the university. Harvard University operates with a mission to institutionalize best practices in sustainable operations and translate research and teaching into practice by using the campus to pilot innovative solutions that can be widely replicated (Harvard Office for Sustainability 2016). HGSE, which is celebrating its centennial year in 2020, resides within the greater institutional framework of Harvard University.

Third, HGSE provides a space where educators from multiple sectors can convene to generate knowledge that improves learning outcomes (Harvard Graduate School of Education 2020a, b), allowing the ideal mix of students who would be taking a curriculum offered at HGSE.

6.3.1 Methodology

Existing literature points out the fruitful and productive partnerships of grassroots initiatives with large scale institutions and/or government organizations (Reeves et al. 2014; Fadeeva et al. 2014). Wi (2019) lays out a 6-stage process of interaction between grassroots organizations and government agencies within a cycle of a climate change education policy adoption. We have adapted this framework to lay out the steps of the co-creation and collaboration of student and faculty within the cycle of a climate change curriculum implementation. Figure 6.1 juxtaposes the adapted 6 stages with the initial one.

The six stages include:

Stage 1: Student Involvement: The Role of Grassroots Innovation
Stage 2: Faculty-Student Exchange: Interaction Zone
Stage 3: Faculty Meeting: The Role of the Institution
Stage 4: Faculty Planning: The Role of the Institution
Stage 5: Implementation: Interaction Zone
Stage 6: Feedback: Interaction Zone

	Bottom Up Approach			Top Down Approach		
	Role of Grassroots	Interaction Zone	Role of Institution	Role of Institution	Interaction Zone	Interaction Zone
	Stage 1	Stage 2	Stage 3	Stage 4	Stage 5	Stage 6
Adapted version for HGSE	Student Involvement	Faculty-Student Exchange	Faculty Meeting	Faculty Planning	Implementation	Feedback
Wi (2019)	Involvement (Initiatives and Activities by Public Education Programs)	Feedback (Communicate suggestions by community)	Advisory (Institution intervenes and provides help)	Policy Planning (Strategic formulation)	Information & Implementation (Policy is deciphered by individual organizations/ departments)	Consultation (Initiates feedback)

Figure 6.1. A Six Step Process of the Grassroots Approach Framework.

The grassroots approach (GRA) to climate change education (bottom row) and the adapted version of the 6-step process (highlighted).

Fig. 6.1 A process for co-creation of a climate change curriculum between university students and faculty

A. Stage 1. Student Involvement (Role of Grassroots)

The ultimate goal of the course is to produce graduates from schools of education worldwide who can serve as leaders in the twenty-first century global movement on the pressing issue of climate change mitigation, adaptation, impact reduction and early warning.

The theory of action on which this proposal is based is:

If a curriculum on climate change is implemented and offered for a minimum of 3 years at HGSE **and** a substantial number of students show interest in enrolling, **then** the course will produce numerous graduates active in various education sectors that have the knowledge, skills, and attitude to help mitigate climate change.

If some HGSE graduates become transformative leaders in education **and** the innovative teaching solutions can be widely replicated (Harvard Office for Sustainability 2016), **then** the course can impart knowledge, skills, and values to the broader community and serve as the blueprint for other education schools.

More specifically, our curriculum set out to attempt the following objectives:

- First, to help students develop knowledge on climate change and the role of different education sectors and educators in overcoming the issue.
- Second, to ingrain the value of being active agents of change, planning and acting to mitigate climate change.

- Third, to build skills and competencies in collaboratively solving problems related to climate change through education.

B. Stage 2. Faculty-Student Exchange (Interaction Zone)
Previous efforts to reorient teacher education to address sustainability and innovations related to ESD were carried out on a personal scale and were short-lived, unable to last beyond personnel and changes in innovation (Hopkins and McKeown 2014). It became evident that our curriculum would also need sustainable support from the institution.

We approached a professor at HGSE, a cognitive scientist whose work centers around ecology, climate change, and the use of virtual and augmented reality in the classroom. In particular, her research identifies ways in which understandings about the nature of causality impact our ability to deal with complexity in our world. Her extensive expertise and experience on the topic made her our top candidate as we approached her with the idea of implementing a climate change curriculum at HGSE.

This exchange indicated that for a curriculum to be introduced at HGSE, we would have to identify and demonstrate the need for a course on climate change. If and only then, faculty could move forward with the motion, and the course would be examined for approval. To aid that process, we administered a needs assessment to demonstrate the existing gaps between current course offerings and student demand for the intended curriculum. Data we collected could paint a better picture of the status quo.

i. Survey Administration and Interpretation
In order to assess preliminary needs and interest in a climate change curriculum for the twenty-first century at HGSE, we administered a Knowledge, Attitude, and Practice (KAP) survey to 66 respondents comprised mostly of HGSE students. A randomized sample was collected through an online survey soliciting students within HGSE and the greater Harvard community. Limitations of the survey include probable bias towards climate activism due to its opt-in design, as well as its small sample size, representing less than 10% of the overall HGSE student population. The 66 participants were affiliated with various programs at HGSE and planned to go into different sectors in education, reflecting the diverse student composition in education sectors that our curriculum targets. Each question was designed specifically to correspond with a Knowledge, Attitude, or Practice (KAP) assessment of student's understanding and interest of climate change. (Table 6.1).

a. Knowledge
A survey of the participants' self-assessed knowledge on climate change revealed 66.2% of the respondents reporting they had gathered some information about the subject, and 15.4% claiming to have very limited knowledge on climate change. To test the actual knowledge of climate change, two multiple-choice questions were administered, questioning the causes and effects of climate change. Only 13 respondents, comprising 19.7% of the total sample answered correctly on both questions,

Table 6.1 KAP survey to understand HGSE student's interests of climate change

	Question	Respondents
Affiliation	State your school affiliation	Among 66 graduate students within the wider Harvard network, 94% of the sample (62) were comprised of students at the Harvard Graduate School of Education
	Which education sector do you plan on going into after graduation?	(33.3%) Non Profit Organization (17.5%) K-12 School System (12.7%) Media/Technology (12.7%) Consulting (7.9%) Government (6%) Higher Education (9.9%) Other
Knowledge	How would you rate your knowledge about climate change? Self-assessed knowledge on ESD	(66.2%) I have gathered some information about the subject. (16.9%) I have been studying the effects of climate change (15.4%) Very limited; I have heard about it, but I am not aware of the fact (1.5%) I am an expert on climate change
	What do you think is the cause of climate change?	22 out of 66 respondents (33.3%) replied correctly
	What are some of the direct effects of global warming?	29 out of 66 respondents (43.9%) answered correctly *13 respondents (19.7%) answered correct for both questions assessing current knowledge of climate change
Attitude	Do you think climate change is happening?	(90.5%) Very certain (7.9%) Somewhat certain (1.6%) Not very certain
	How concerned are you about climate change?	(64.6%) Very concerned (30.8%) Somewhat concerned
	How big of a role do you think *education* plays in mitigating climate change?	(64.6%) Very big (30.8%) Somewhat big (3.1%) Not very big (1.5%) Not at all big
	How big of a role do you think *educators* plays in mitigating climate change?	(49.2%) Very big (35.4%) Somewhat big (13.8%) Not very big (1.5%) Not at all big

(continued)

Table 6.1 (continued)

	Question	Respondents
Practice	What are some efforts you've undertaken to mitigate climate change?	(80.7%) Reducing one's own carbon footprint (including taking public transportation, recycling, reducing plastic use, buying sustainable products, limiting the consumption of meat, etc.) (11.5%) Raising awareness (sharing information with families and talking to coworkers regarding the issue etc.) (7.7%) Engaging in political action (voting for leaders who demonstrate concern for climate change, putting pressure on leadership to encourage compost, and joining climate strikes) (5.8%) No effort (3.8%) Skepticism of individual efforts in mitigating climate change
	What are some efforts you've undertaken to mitigate climate change *through education*?	(25.5%) Informal conversations (23.4%) Teaching students (6.4%) Using social media (6.4%) Organizing events (6.4%) Advocacy (6.4%) No attempts
Interest	Would you be interested in taking a course at HGSE on climate change and education?	(52.3%) Yes (47.7%) No

demonstrating a significant gap in students' *actual* knowledge of climate change.

b. Attitudes

   90.5% of the respondents were very certain that climate change was happening, showing a strong consensus on the existence of the phenomenon. 95.4% of respondents expressed concerns regarding climate change, with 64.6% of the respondents responding that they were very concerned about climate change. There was a general consensus with regards to whether *education* had a significant role in mitigating climate change, with 95.4% of respondents selecting 4 or 5 on a scale of 5, with higher numbers indicating greater significance. Strong agreement (84.6% of respondents who answered a 4 or 5 on a scale of 5) was also observed in responses to the question of whether *educators* were significant in mitigating climate change.

c. Practice

   By analyzing qualitative responses to questions asking one's efforts taken to mitigate climate change, we could extract five general themes that represented the responses: reducing one's own carbon footprint, raising awareness, engaging in political action, no effort, and skepticism of individual effort in mitigating climate change. The majority of the respondents (90.4%) have made efforts in mitigating climate change, with the most common responses related to reducing their own carbon footprint (80.7%).

Next, an analysis of one's attempt to mitigate climate change through education yielded six different themes: informal conversations, teaching students, using social media, organizing events, and advocacy. The most common attempts were engaging in informal conversations (25.5%) and teaching students on the topic of climate change (23.4%).

d. Interest

Thirty-four students, or 52.3% of the total number of respondents, showed an interest in taking a course on climate change and education.

Whereas needs assessments like the one we conducted are not meant to be generalizable nor predictive of actual behavior, this figure provides some indication that there is considerable interest in CCE among students at the Harvard Graduate School of Education. We shared the results of the survey with the academic leadership of the school, and the feedback was immediately positive. The urgency of the situation and the demand from the student population clearly demonstrated the need for this class.

C. Stage 3. Faculty Meeting (Role of Institution)

Creating new classes can be a time-consuming effort that requires several layers of review within the faculty and by a governing body that controls curricular and academic changes. Such a review process typically requires creation of a syllabus and reading list (Appendix A) for the course as well as examining the credentials and expertise of those who would teach the course to show they have adequate expertise.

We waited for a follow-up from faculty, and much to our delight, were informed that such a course was approved.

D. Stage 4. Faculty Planning (Role of Institution)

Each institution charts out a curriculum implementation process, systematically organizing what will be taught, who will be taught, and how it will be taught. At HGSE, there is a three-step process that aids faculty in planning a course curriculum.

- Phase 1: One on One Conversation
- Phase 2: Course Preparation
- Phase 3: Teaching Team Meeting.

i. Phase 1: One on One Conversation

In this phase, faculty reviews the course planning guide and forms a teaching team. Together, they discuss and agree on due dates.

ii. Phase 2: Course Preparation

In the course preparation phase, design elements of the course are carefully considered. The scope (breadth of knowledge, skills, attitudes, and behaviors) and the sequence (order) of the course would be discussed. Appropriate materials, pedagogy, and effective class activities are chosen within the design of the curriculum (Brundiers et al. 2014). Formative and summative evaluation methods are examined which will measure the effectiveness of the curriculum.

Faculty members are supported by the Harvard Graduate School Course Planning Team. The Teaching and Learning Lab provides faculty support through individual consultations and group professional opportunities to craft intelligent and thoughtful course designs as well as producing instructional resources (HGSE Teaching and Learning Lab 2020).

iii. Phase 3: Tea+ching Team Meeting

Faculty assistants, faculty, and teaching fellows are determined and meet to discuss the facilitation of the class.

E. Stage 5. Implementation (Interaction Zone)

The authors participated in the faculty driven process of the implementation by offering assistance and support in the ideation and design of an ESD curriculum aimed at the school of education. Two prototype lessons of the curriculum can be accessed in the Appendix B.

i. Standards

Our curriculum seeks to meet standards provided by the 2030 Sustainable Development Goal, target 13 Climate Action. In particular, goal 13.3 focuses on improving education, awareness-raising, and building institutional capacity related to climate change mitigation, adaptation, impact reduction and early warning. Furthermore, we have aligned our curriculum to accommodate the four pillars of HGSE foundational elements of Learning, Development, and Teaching; Evidence; Equity and Opportunity; Organizations and Systems (Harvard Graduate School of Education 2020a, b). In order to emphasize the importance of making knowledgeable behavioral decisions, we included the five basic competencies as defined by Wiek et al. (2011): Systems thinking competency, Anticipatory competency, Normative competency, Strategic competency, and Interpersonal competency.

ii. Format

Climate Change for Educators will be a 12-week semester long course, comprised of two 90-min classes focusing on a weekly theme. We based the overarching framework of the curriculum design using the UN SDG aligned curriculum *Empowering Students to Improve the World in Sixty Lessons* (Reimers 2017) in addition to *Becoming Global Thinkers: Thinking about Distant Causes and Effects, Causal Learning in the Classroom (CLIC)* (Grotzer et al. 2015).

Furthermore, our lesson plan prototype has been designed so that each of the weekly topics and units can function independently of one another. Thus, they can be separated and blended into existing structures of curricula units to impart knowledge, skills, and values that teach educators about the critical issue of climate change.

iii. Pedagogical Design

a. Knowledge Transfer

In 1956, Bloom categorized cognitive learning objectives in a progressive hierarchy from least to most complex levels which include: knowledge, comprehension, application, analysis, synthesis, and evaluation. Based on his taxonomy, this curriculum devotes the first class of each week building upon basic, fundamental skills such as 'knowledge,' and 'comprehension' via lectures and engaging discussions. A knowledge-based start will help to foster an atmosphere of intellectual discourse in the classroom (Fook 2012), and specifically of accumulating, deepening, and transferring knowledge of the environment to a large mass of educators (Radaković et al. 2017).

After a short lecture, faculty will engage students in a lively discussion, which will foster student engagement.

b. Engagement
Lectures are followed by an invigorating student led discussion, for students to 'synthesize' (Bloom 1956) what they have learned. Discussions help bring out the importance of engagement and fostering a culture of student-centered learning (Anderson 2012). Discussion prompts will center around the essential questions posed by faculty, and this active facilitation is enhanced as faculty demonstrate the changing role of a teacher; from "content expert" to "curriculum facilitator," in this new era of learning (Godsey 2015). Active listening is encouraged, and main points are organized on a board framing the argument and building insights.

c. Cross-Disciplinary
Climate change education inevitably requires having to incorporate a blend of multi-disciplinary academic subjects (Lindblom-Ylänne et al. 2006; Lueddeke 2003; Nevgi et al. 2004; Singer 1996). Topics within the scope of this curriculum include socio-political issues surrounding the scientific facts of climate change, innovative technology serving as possible solutions, as well as the role of education in mitigating climate change. Week 3 will touch upon the cognitive function of how people learn, with regards to understanding that climate change requires an 'action-at-a-distance' approach. Weeks 4–11 will cover a myriad of ways in which education can help mitigate the impact of climate change. Some topics include curriculum design, professional development of teachers, school operations, informal education, measurements, education policy, and climate justice. In the final week of class, students will give their presentations and share the takeaways from the course.

d. Project Based
The second class each week will consist of 'applying,' 'synthesizing,' and 'evaluating' (Bloom 1956) what they know via engaging projects. Numerous studies have highlighted the benefits of active, project-based learning (Leigh 2009). The semester-long project involves a group of 3–5 students (Henke 1985) with varying levels of experience in multiple education sectors, collaborating to craft a holistic educational strategy to mitigate climate change. Students are asked to formulate project groups around a jurisdiction with meaningful personal ties, helping to contextualize learning as local, tangible, and personally relevant (Cone et al. 2012; Anderson 2012). Each jurisdiction will highlight the different agenda and perspectives present today in the twenty-first century global climate change movement. Students will give presentations in the final week and

submit a 20-page-paper as a final project. Throughout the project, students are active participants in their own learning which will include the design of their experiences and the realization of their learning outcomes. Ultimately, students can take full ownership of their own learning. This class will involve weekly group assignments (Monroe et al. 2017) that students will start in the class but finish as collaborative homework.

e. Situated
The second half of the week additionally focuses on student-centered activities, guest lecturers, simulations, and excursions to provide a thought-provoking experience for students to experience real life perspectives in the topic of climate change. In week 7, students will take an excursion to a nearby environmental institution. With multiple studies revealing the importance of making the distant threat of climate change personally relevant and meaningful (Shome et al. 2009; Fook 2012; Moser and Dilling 2007; Wibeck 2014), excursions can help make the threat of climate change real, tangible and immediate (Cone et al. 2012). We also bring in local guest speakers committed to the field to bring expert knowledge into the classroom (Leigh 2009; Theobald et al. 2015) serving as a relevant local source of inspiration to the students all the while minimizing carbon footprint expenditure.

f. Real World Situations
Our audience are adult learners, who see themselves as capable of self-direction and incentivized by tasks that will prepare for social and occupational role competency (El Sawi 1996). Each week, we provide assignments that carry out practical exercises that encourage learners to put into practice the theories they learned. Our adult learners have the autonomy to carry out their given task, on very practical elements that can be utilized in the workforce.

g. Faculty Involvement
Students will be encouraged to meet with instructors at least once every 3 weeks so that expectations and standards from both sides of the teaching team and the students are well integrated, coherent, and harmonized. Intimate feedback from the teaching team enhances student learning, and the instructor takes away with a solid knowledge and understanding of student's progress.

h. Accountability
Students will be assessed on a weekly basis, with weekly projects consisting 60% of the total grade, and the semester long final projects being worth 40%. As students are working in groups, accountability measures such as self-assessments and peer-evaluation sheets will be collected by every member of the group. Student projects would be assessed via a rubric.

F. Stage 6 Feedback
In this final stage, student-faculty interaction is oriented towards the students who are taking the course and the faculty teaching the course. Feedback from

students in the form of formative and summative evaluations is collected to improve and sustain the curriculum.

6.4 Implications & Conclusion

Our chapter described and analyzed efforts of implementing a climate change curriculum within the Harvard Graduate School of Education (HGSE) that helps to build competencies for potential leaders in different education sectors so that they can combat climate change collaboratively. However, our goal is not the development and implementation of a single curriculum. We aspire to achieve a collective legacy of preparing educators within the whole-of-system and achieve widespread change, which will require concentrated and concerted efforts to disseminate the importance of climate change curricula at multiple schools of education.

There are obvious limitations in focusing solely on a curriculum without simultaneous efforts to promote sustainability as endorsed within a whole-school model of ESD within the entire Harvard Graduate School of Education. Moreover, the disruption of Covid-19 may have significant, damaging effects to the immediate implementation of the curriculum, as well as its aftermaths in the upcoming years should there be a shift to a digital model of education. Furthermore, the institution shift into a mode to mitigate the impact of the pandemic on the school may induce several entanglements previously unanticipated in the conception of the curriculum.

To answer our original question: If students played an equal role in the development of curricula, selection of course content, and initiation of applied projects, how different might the impact of the academy become?

Our curriculum explored the successful beginnings of a student/faculty co-creation and collaboration. This is not to say there were not tensions between expert and novice and between discovery and direction. For a successful blend of bottom-up and top-down interaction, multiple components must come together under mutual respect, common understanding, and shared responsibility. Learner agency, increased satisfaction with the academy, and heightened engagement came with the author's participation of a curriculum design. The institution was provided with insights into students' perspective and needs as well as enhanced capacity to promote a holistic engagement with learners.

Ultimately, a whole-of-systems approach to climate change means that the entirety of the system promotes action for reducing climate change, sharing the responsibility within all. While each individual component of these five perspectives is integral to creating a culture of change, it is the sum of the respective perspectives and the interaction between all of the elements that will be critical in the 21st global movement of shared environmental responsibility. And a curriculum like ours, educating future leaders within schools of education, is a step closer in the right direction of attaining a sustainable future for all.

Appendices

Appendix A. Syllabus

Class title: Climate change for educators

	Lecture classes	Project based classes	
Week 1	Class 1 Introduction to climate change education	Class 2 Activity #1: Formation of project groups.	Building up a background knowledge of climate change
Week 2	Class 3 Socio-political issues surrounding the scientific facts of climate change	Class 4 Guest lecturer on solar engineering Activity #2: Innovative technology	
Week 3	Class 5 What makes climate change hard to accept?	Class 6 Guest lecturer EcoMuve Activity #3: Ethics and individual impact	
Week 4	Class 7 Whole-of system model to climate change education	Class 8 Guest lecturer: GSD professor Activity # 4: Identifying sustainable school operations	Climate change within local school systems
Week 5	Class 9 Supporting teachers to teach	Class 10 Activity # 5: Designing ProD activity and teacher support systems	
Week 6	Class 11 What makes for a good climate change curriculum?	Class 12 Activity #6: Creating a curriculum based on competencies and best learning practices	
Week 7	Class 13 Informal education – Educating the mass	Class 14 Excursion. Climate change in the real world. Visit the Harvard Museum of Natural History: Climate change exhibition. Activity # 7: Design an awareness campaign to draw in the audience.	
Week 8	Class 15 Economic metrics, measurements, and evaluations	Class 16 Activity # 8: Find measurements of HGSE operations and assess its energy consumption.	

(continued)

	Lecture classes	Project based classes	
Week 9	Class 17 Climate change education in the United States	Class 18 Activity # 9: Write a policy memo	Climate change in the world
Week 10	Class 19 The state of global climate change education	Class 20 Climate change simulation – world climate	
Week 11	Class 21 Climate justice and ethics: Migration, resilience, and adaptation	Class 22 Review class Activity # 10: Prepare for a 10-min final presentation	
Week 12	Class 23 Presentations of projects &wrap up.	Class 24 Presentations of projects &wrap up. Activity # 11: Final paper	Review

Appendix B. Prototype Lessons of a CCE Curriculum at HGSE

Week 1 Class 1

Introduction to Climate Change Education

Knowledge & Application

Time: 90 min

Standards: Sustainable Development Goals 13.1, 13.3; HGSE foundational element: Learning, Development, and Teaching; Evidence; Equity and Opportunity; Organizations and Systems

Competencies: Systems-thinking, Anticipatory, Normative Strategic, Interpersonal

Summary/Rationale:

The 2030 Agenda for Sustainable Development, adopted by all United Nations Member States in 2015, provides a shared blueprint for peace and prosperity for people and the planet.

Sustainable Development Goal 13 asks member states to take urgent action to combat climate change and its impacts. Specifically, target 13.3 implores nations worldwide to improve education, awareness-raising, and human and institutional capacity on climate change mitigation, adaptation, impact reduction and early warning. Achieving this legacy goal requires involving various educators, agents in delivering education and raising awareness. This course will help explore how a climate change curriculum at the Harvard Graduate School of Education can prepare educators to mitigate climate change.

Instructional Goal: Students are provided with a general overview of climate change education and the important role of educators within education for sustainable development.

Student Learning Objectives

- Cognitive:
  - The learner understands that the anthropogenic emission of greenhouse gasses account for almost all the increase in greenhouse gasses in the atmosphere over the last 150 years.
  - The learner understands the role educators can assume in the shared responsibility of environmental sustainability.
  - The learner knows the five competencies required in climate change education.
- Socio-Emotional
  - The learner can recognize that the protection of the global climate is an essential task for everyone and the need to re-evaluate our worldview and everyday behaviors.
  - The learner understands the importance of a whole-of-systems model, in its interaction and support.
- Behavioral Learning
  - The learner is able to communicate and endorse positive approaches to CCE.

Essential questions:

- *How does the role of education and educators fit in the global movement of climate change?*
- *What is climate change and its causes? How does human play a part in climate change?*
- *What is the purpose of climate change education (CCE)?*
- *How can we approach CCE so that knowledge can be transpired into action?*

Sequence of activities

1. (30 min) Introduce the course

Faculty welcomes students and introduces themselves. He/She will then go over the course format, pedagogical choices, and the scope of the curriculum. Faculty will model best practices of teaching and learning both in pedagogy and content.

Something you might say," I will go over course logistics and sketch out the main issues in a very preliminary way. Students(you) should feel free to raise questions of each other or of me."

2. (35 min) Lecture – Knowledge transfer

Faculty may want to introduce a thought-provoking video on CC or state of CCE to spark this lecture. Briefly explain the key points of the readings students have been asked to complete for homework.

(Video: National Geographic https://www.youtube.com/watch?v=QwLyscT3NgI)

(a) Introduction to Climate Change
Brief overview of history of Earth's atmosphere: what is a greenhouse gas?
Students should have completed the following readings for homework:
Reading: Allégre & Schneider, Scientific American, 2005. The evolution of Earth
Reading (optional): Meteorology Today, Chapter 18, Earth's Changing Climate

(b) Climate Change and Education – Importance of Educating Educators Reading: https://oxfordre.com/environmentalscience/view/10.1093/acrefore/9780199389414.001.0001/acrefore-9780199389414-e-56

(c) What are Competencies in CCE and why is it important? What are some tensions seen in CCE today?
Reading: Vega-Marcote, Pedro & Varela-Losada, Mercedes. (2016). Basic Teacher Training Oriented Toward Sustainability: Why and How to Carry It Out Today?

(d) A Whole-of Systems model in CCE/ESD.
Reading: Ferreira, Jo-Anne & Ryan, Lisa & Tilbury, Daniella. (2007). Mainstreaming education for sustainable development in initial teacher education in Australia: A review of existing professional development models. Journal of Education for Teaching.

3. (25 min) Discussion – Applying what you know.

This last 25 min is reserved for a whole-class reflection. Active, engaged student participation is important to maximize learning outcomes.

Essential question: How does the role of education and educators fit in the global movement of climate change?

There are several ways you can set up this discussion.

1. A whole-class discussion in which discussions are freely encouraged
2. A case study discussion – Encourage students to think about the prompt by providing a specific case of CCE. As students discuss the prompt, organize their ideas on the board through a framework to bring out further insights and innovative approaches.
3. Negotiation/Simulation – Provide students with a designated 'role.' i.e. teacher, student, administration, government etc. Allow students to explore this discussion through the lens of the role they are given.

Discussion should be primarily student led, with faculty facilitating the discussion by bringing out diverse voices (with the occasional cold calling) and probing students to capture relevant knowledge and nuances of the case.

Week 1 Class 2
Activity #1 Formation of Groups
Application & Evaluation
Time: 90 min
Project

Standards: Sustainable Development Goals 13.2 HGSE foundational element: Learning, Development, and Teaching; Organizations and Systems

Competencies: Systems-thinking, Interpersonal, Normative, Anticipatory

Summary/Rationale:

Every 2nd half of the week, classes are centered around a semester long project, involving a group of 3–5 students (Henke 1985) with varying levels of experience in multiple education sectors, collaborating to craft a holistic educational solution to mitigate climate change.

Students will formulate their own groups based on personal interests and goals, but the groups must be centered around a common geographical tie. The jurisdiction of their choice will highlight the different agendas and perspectives present today in the twenty-first century global movement of sustainable development. As students pick a jurisdiction with meaningful personal ties, this helps to contextualize climate change learning as local, tangible, and personally relevant (Cone et al. 2012; Anderson 2012).

Each week, groups are given structured activities that allow members to explore a different perspective of CCE, using systems thinking to approach the multi-layered movement. Through class presentations and group assignments there will be ample opportunities for peer-learning and cultivating global citizenship. Students will give presentations in the final week and submit a 20-page-paper as their final project.

Notes: Make sure students understand "group dynamics" and go over the "key mistakes of forming groups" slide. It is important for the class to understand how to select group members and to function well together throughout the semester.

Student Learning Objectives

- Cognitive
  - The learner can discern that different jurisdictions face different dilemmas concerning climate change.
- Socio-emotional
  - The learner is able to encourage others to protect the climate.
  - The learner learns to work in a group setting.
- Behavioral
  - The learner engages in cooperative team dynamics.

Sequence of activities

1. (5 min) Introduce the activity of the day

Explain to the class the expectations for classes in the 2nd half of the week. We will establish norms for this 'project-based activities' class.

Introduce the semester long project by starting with the goal and objective. Their primary goal for this class is to form project groups they will be working with for the remainder of the semester.

2. (30 min) Step-by-step approach to applying concepts into projects.

Step 1. How does 'geography' play a factor in climate change education? How does the country/region/political affiliation/organization you belong to/identify with shape your perspective on climate change/environment?

Reading Assignment:

https://www.climateinteractive.org/programs/world-climate/facilitator-resources/

Skim the 6 briefing statements of: The United States, European Union, Other Developed, China, India, Other Developing.

PowerPoint slide: How might geography/culture influence my view?

Developed Nation: (Pro Climate Change) European Union, Canada, Japan, South Korea (Majority of OECD member states)
Developed Nation: United States (Political affiliations in the United States)
Rapidly Emerging Nations: China, India
Developing Nation: Ghana, Myanmar
Pacific Island Nation: Maldives

Allow students to take 5 min to reflect on the jurisdiction they would like to focus on throughout this semester. Make sure students are aware that the groups they choose in this class may be difficult to change throughout the semester.

Step 2. Working in Groups:

Reading Assignment-

Edmondson, A. C. (2012) Teaming to learn, innovate, and compete. In Teaming: How organizations learn, innovate, and compete in the knowledge economy. San Francisco: Jossey-Bass. Chapter 2, pp. 45–80

PowerPoint Slide-

What are some common misconceptions to forming groups?

1. More is better, so make the team big → Optimal group size: 4.6
2. Similar people will get along better → Heterogeneity: variety of skills and backgrounds works better
3. Everyone understands how to work in a group → Not so! You must develop shared understanding of how to work in a group; Don't forget to keep looking outward – for new sources, new inspirations; Don't ignore an "off-target" comment … it may reflect an essential perspective from a different group; Teamwork

can be stressful: "Serious work means serious tensions"; Fear of not getting along, Fear of being "wrong," Silence is easier than speaking

What are positive team behaviors?

1. Everyone has an obligation to participate
2. From brainstorming forward, build on others' good ideas
3. Respond honestly – doesn't help the group to withhold what you really think
4. Critique ideas; not individuals – make the team a psychologically "safe space"
5. Be flexible
6. Seek consensus ... ... and have a process for making decisions
7. Share leadership ... ... but define clear roles for specific tasks and stages
8. Minimize egos ... ... and maximize fun and humor'

Conduct an online survey- Ask students to complete a brief survey of one's strengths, interests, and competencies.

Results will be displayed on the screen, and the class can see the unique skillsets each members of the class possess.

Step 3. Formulate your team.

*3–5 members to a team based on regional interest, and skills assessment, students are asked to move around to find members.

3. (30 min) Meeting the Team

Faculty and TFs go around to make sure groups are working well.

The main task is allowing students to formulate their groups for the semester. Once small groups begin to congregate, facilitate the session by asking members to introduce themselves, their interest in a geographic sector and for this course overall.

*Note that students will be adding/dropping the course until the add/drop deadline.

Assignment: Write up a short reflection about:

1. What are some group norms you tend to establish, follow, and promote?
2. Why you are interested in this particular jurisdiction?
3. What you hope to learn and what questions you hope to answer.

You are welcome to write a few short paragraphs, or draw up an outline, or list a few bullet points. The goal is to set a brief agenda for the semester, generate ideas, and establish expectations.

While the first assignment is an individual reflection, the majority of assignments are group driven, and students are encouraged to meet outside of class time for the purpose of the projects.

References

Anderson, A. (2012). Climate change education for mitigation and adaptation. *Journal of Education for Sustainable Development, 6*(2), 191–206.

Aspen Global Change Institute. (2019). *Fundamentals*. Advancing global change science and solutions. Retrieved from https://www.agci.org/education/fundamentals

Bloom, B. S. (1956). Taxonomy of educational objectives: The classification of educational goals; Handbook I: Cognitive domain. In M. D. Engelhart, E. J. Furst, W. H. Hill, & D. R. Krathwohl (Eds.), *Taxonomy of educational objectives: The classification of educational goals; Handbook I: Cognitive domain*. New York: David McKay.

Brundiers, K., Savage, E., Mannell, S., Lang, D. J., & Wiek, A. (2014). Educating sustainability change agents by design: Appraisals of the transformative role of higher education. In Z. Fadeeva, L. Galkute, C. Mader, & G. Scott (Eds.), *Sustainable development and quality assurance in higher education: Transformation of learning and society* (pp. 196–229). New York: Palgrave Macmillan. https://doi.org/10.1057/9781137459145.

Christens, B. D. (2010). Public relationship building in grassroots community organizing: Relational intervention for individual and systems change. *Journal of Community Psychology, 38*(7), 886–900.

Clements, K., Dimenstein, L., Melampy, E., Pollack, J., Simon, S., Song, Q., Spitz, S., Weatherall, G., Westrate, R., & Wriston, E. (2020). *Regulatory rollback tracker.* Environmental and Energy law program. Retrieved from https://eelp.law.harvard.edu/regulatory-rollback-tracker/

Cone, J., Rowe, S., Borberg, J., & Goodwin, B. (2012). Community planning for climate change: Visible thinking tools facilitate shared understanding. *Journal of Community Engagement and Scholarship, 5*(2), 7–19.

Cook-Sather, A., Bovill, C., & Felten, P. (2014). *Engaging students as partners in learning and teaching: A guide for faculty*. San Francisco: Jossey Bass.

de Souza, R. T. (2007). *NGOs and empowerment: Creating communicative spaces in the realm of HIV/AIDS in India* [Doctoral dissertation, Purdue University]. ResearchGate.

Denby, L., & Rickards, S. (2016). An approach to embedding sustainability into undergraduate curriculum: Macquarie university, Australia case study. In W. Leal Filho & P. Pace (Eds.), *Teaching education for sustainable development at university level* (pp. 9–33). Dordrecht: Springer.

Dunne, E. (2016). Design thinking: A framework for student engagement? A personal view. *Journal of Educational Innovation, Partnership and Change, 2*(1). https://doi-org.ezp-prod1.hul.harvard.edu/10.21100/jeipc.v2i1.317

Dyer, G., & Andrews, J. (2011). *Higher education's role in adapting to a changing climate*. Boston: American College & University Presidents' Climate Commitment.

El Sawi, G. (1996). *Population education for non-formal education programs of out-of-school rural youth*. Food and Agriculture Organization of the United Nation. Retrieved from http://www.fao.org/3/ah650e/AH650E00.htm

Environmental Defense Fund & Meister Consultants Group. (2017). *Now hiring: The growth of America's clean energy & sustainability jobs*. Environmental Defense Fund Climate Corps. Retrieved from http://edfclimatecorps.org/sites/edfclimatecorps.org/files/casestudy/the_growth_of_americas_clean_energy_and_sustainability_jobs.pdf

Fadeeva, Z., Galkute, L., Mader, C., & Scott, G. (2014). Assessment for transformation: Higher education thrives in redefining quality systems. In Z. Fadeeva, L. Galkute, C. Mader, & G. Scott (Eds.), *Sustainable development and quality assurance in higher education: Transformation of learning and society* (pp. 36–57). New York: Palgrave Macmillan. https://doi.org/10.1057/9781137459145.

Fahey, S. J. (2012). Curriculum change and climate change: Inside outside pressures in higher education. *Journal of Curriculum Studies, 44*(5), 703–722. https://doi.org/10.1080/00220272.2012.679011.

Ferreira, J. A., Ryan, L., & Tilbury, D. (2007). Mainstreaming education for sustainable development in initial teacher education in Australia: A review of existing professional development models. *Journal of Education for Teaching, 33*(2), 225–239.

Fisher, J. (1998). *NGOs and the political development of the third world.* West Hartford: Kumarian Press.

Fook, C. Y. (2012). Best practices of teaching in higher education in United States: A case study. *Procedia - Social and Behavioral Sciences, 46*, 4817–4821. https://doi.org/10.1016/j.sbspro.2012.06.341.

Forbes, C. T., & Zint, M. (2010). Elementary teachers' beliefs about, perceived competencies for, and reported use of scientific inquiry to promote student learning about and for the environment. *The Journal of Environmental Education, 42*(1), 30–42.

Fraser, E. D., Dougill, A. J., Mabee, W. E., Reed, M., & McAlpine, P. (2006). Bottom up and top down: Analysis of participatory processes for sustainability indicator identification as a pathway to community empowerment and sustainable environmental management. *Journal of Environmental Management, 78*(2), 114–127.

García Bacete, F. J., Ferrá, P., Monjas, M. I., & Marande, G. (2014). Las relaciones del profesoradocon el alumnado en aulas del ciclo inicial de Educación Primaria. *Adaptación del questionnaire on Teacher Interaction-Early Primary (QTI-EP), 19*(1), 211–231.

Godsey, M. (2015, March 25). The deconstruction of the K-12 teacher: When kids can get their lessons from the Internet, what's left for classroom instructors to do?. *The Atlantic*. Retrieved from https://www.theatlantic.com/education/archive/2015/03/the-deconstruction-of-the-k-12-teacher/388631/

Grotzer, T. A., Derbiszewska, K., Gramling, M. D., & Solis, L. S. (2015). *Becoming global thinkers: Thinking about distant causes and effects, causal learning in the classroom (CLIC) modules.* Cambridge, MA: President and Fellows of Harvard College.

Harker-Schuch, I. (2019). Why is early adolescence so pivotal in the climate change communication and education arena? In W. Leal Filho & S. L. Hemstock (Eds.), *Climate change and the role of education* (pp. 279–290). Cham: Springer International Publishing.

Harvard Graduate School of Education. (2020a). *About the Harvard Graduate School of Education.* Retrieved from https://www.gse.harvard.edu/about

Harvard Graduate School of Education. (2020b). *Master's 2021: FAQs.* https://www.gse.harvard.edu/masters2021/faq

Harvard Office for Sustainability. (2016). *Our plan.* Sustainability at Harvard. Retrieved from https://green.harvard.edu/campaign/our-plan

Henke, J. W., Jr. (1985). Bringing reality to the introductory marketing student. *Journal of Marketing Education, 7*(3), 59–71.

Hess, D., & Collins, B. (2018). Climate change and higher education: Assessing factors that affect curriculum requirements. *Journal of Cleaner Production, 170*, 1451–1458. https://doi.org/10.1016/j.jclepro.2017.09.215.

HGSE Teaching and Learning Lab. (2020). *Services.* Retrieved from https://tll.gse.harvard.edu/services

Hopkins, C., & McKeown, R. (2014). *Teacher education and education for sustainable development: Ending the DESD and beginning the GAP.* Retrieved from https://www.kdp.org/initiatives/pdf/TeacherEdESDChairReport.pdf

Hungerford, H. R., & Volk, T. L. (1990). Changing learner behavior through environmental education. *The Journal of Environmental Education, 21*(3), 8–21.

Kasser, T., & Ryan, R. M. (1996). Further examining the American dream: Differential correlates of intrinsic and extrinsic goals. *Personality and Social Psychology Bulletin, 22*, 280–287.

Lambrechts, W., Mulà, I., Ceulemans, K., Molderez, I., & Gaeremynck, V. (2013). The integration of competences for sustainable development in higher education: An analysis of bachelor programs in management. *Journal of Cleaner Production, 48*, 65–73.

Leigh, K. (2009). Energy busters Norfolk schools fight climate change. *Environmental Education, 91*, 13–14.

Lindblom-Ylänne, S., Trigwell, K., Nevgi, A., & Ashwin, P. (2006). How approaches to teaching are affected by discipline and teaching context. *Studies in Higher Education, 31*(03), 285–298.
Lueddeke, G. R. (2003). Professionalising teaching practice in higher education: A study of disciplinary variation and 'teaching-scholarship'. *Studies in Higher Education, 28*(2), 213–228.
Mercer-Mapstone, L., Dvorakova, S. L., Matthews, K. E., Abbot, S., Cheng, B., Felten, P., Knorr, C., Marquis, E., Shammas, R., & Swaim, K. (2017). A systemic literature review of students as partners in higher education. *International Journal for Students as Partners, 1*(1), 1–23. Retrieved from https://www.researchgate.net/publication/316764140_
Molthan-Hill, P., Worsfold, N., Nagy, G., Leal Filho, W., & Mifsud, M. (2019). Climate change education for universities: A conceptual framework from an international study. *Journal of Cleaner Production, 226*, 1092–1101.
Monroe, M., Plate, R. R., Oxarart, A., Bowers, A., & Chaves, W. A. (2017). Identifying effective climate change education strategies: A systematic review of the research. *Environmental Education Research, 25*(6), 791–812. https://doi.org/10.1080/13504622.2017.1360842.
Moser, S. C., & Dilling, L. (2007). Toward the social tipping point: Creating a climate for change. In S. C. Moser & L. Dilling (Eds.), *Creating a climate for change: Communicating climate change and facilitating social change* (pp. 491–516). Cambridge: Cambridge University Press.
Nevgi, A., Postareff, L., & Lindblom-Ylänne, S. (2004). *The effect of discipline on motivational and self-efficacy beliefs and on approaches to teaching of Finnish and English university teachers*. Paper presented at EARLI SIG Higher Education Conference. Retrieved from https://tuhat.helsinki.fi/ws/portalfiles/portal/29807776/Nevgi_etal_boat_18_210604.pdf
Paul, K., & Tan, A. S. (2003). Democracy and the grassroots sector in Singapore. *Space and Polity, 7*(1), 3–20.
People's Association, Singapore [PA]. (2011). *About grassroots organisations*. Retrieved from http://www.pa.gov.sg/about-grassroots-organisations.html
Pompeo, M. R. (2019, November 4). *On the U.S. Withdrawal from the Paris Agreement*. U.S. Department of State. Retrieved from https://www.state.gov/on-the-u-s-withdrawal-from-the-paris-agreement/
Radaković, J. A., Petrović, N., Milenković, N., Stanojević, K., & Đoković, A. (2017). Improving students' higher environmental and climate change knowledge: A case study. *Polish Journal of Environmental Studies, 26*(6), 2711–2719. https://doi.org/10.15244/pjoes/69645.
Räthzel, N., & Uzzell, D. (2009). Transformative environmental education: A collective rehearsal for reality. *Environmental Education Research, 15*(3), 263–277.
Reeves, A., Lemon, M., & Cook, D. (2014). Jump-starting transition? Catalysing grassroots action on climate change. *Energy Efficiency, 7*(1), 115–132.
Reimers, F. M. (2017). *Empowering students to improve the world in sixty lessons: Version 1.0*. North Charleston: CreateSpace Independent Publishing Platform.
Reimers, F. M. (2020). *Educating student to improve the world*. Singapore: Springer. https://doi.org/10.1007/978-981-15-3887-2.
Rickinson, M. (2001). Learners and learning in environmental education: A critical review of the evidence. *Environmental Education Research, 7*(3), 207–320.
Roorda, D. L., Koomen, H., Spi lt, J. T., & Oort, F. J. (2011). The influence of affective teacher-student relationships on students' school engagement and achievement: A meta-analytic approach. *Review of Educational Research, 8*(4), 493–529.
Rothman, J. (1996). The interweaving of community intervention approaches. *Journal of Community Practice, 3*(3–4), 69–99.
Schultz, P. W. (2002). Knowledge, information, and household recycling: Examining the knowledge-deficit model of behavior change. In T. Dietz & P. Stern (Eds.), *New tools for environmental protection: Education, information, and voluntary measures* (pp. 67–82). Washington, DC: National Academy of Sciences.
Shome, D., Marx, S., Appelt, K., Arora, P., Balstad, R., Broad, K., Freedman, A., Handgraaf, M., Hardisty, D., Krantz, D., Leiserowitz, A., LoBuglio, M., Logg, J., Mazhirov, A., Milch, K., Nawi, N., Peterson, N., Soghoian, A., & Weber, E. (2009). *The psychology of climate change*

communication: A guide for scientists, journalists, educators, political aides, and the interested public. New York: Center for Research on Environmental Decisions.

Singer, E. (1996). Espoused teaching paradigms of college faculty. *Research in Higher Education, 37*, 659–679.

Smith, D. H. (2000). *Grassroots associations*. Newbury Park: Sage Publications.

Stephens, J. C., Hernandez, M. E., Román, M., Graham, A. C., & Scholz, R. W. (2008). Higher education as a change agent for sustainability in different cultures and contexts. *International Journal of Sustainability in Higher Education, 9*(3), 317–338.

Stern, M. J., Powell, R. B., & Ardoin, N. M. (2010). Evaluating a constructivist and culturally responsive approach to environmental education for diverse audiences. *Journal of Environmental Education, 42*(2), 109–122.

Stevenson, K., Peterson, M., & Bradshaw, A. (2016). How climate change beliefs among U.S. teachers do and do not translate to students. *PLoS One, 11*(9), e0161462. https://doi.org/10.1371/journal.pone.0161462.

Theobald, E. J., Crowe, A., HilleRisLambers, J., Wenderoth, M. P., & Freeman, S. (2015). Women learn more from local than global examples of the biological impacts of climate change. *Frontiers in Ecology & the Environment, 13*(3), 132–137. https://doi.org/10.1890/140261.

UNESCO. (2014). *UNESCO roadmap for implementing the global action programme on education for sustainable development.* Retrieved from https://sustainabledevelopment.un.org/content/documents/1674unescoroadmap.pdf

UNESCO. (2015). *Not just hot air: Puting climate change education into practice*. Paris: UNESCO.

Van der Leeuw, S., Wiek, A., Harlow, J., & Buizer, J. (2012). How much time do we have? Urgency and rhetoric in sustainability science. *Sustainability Science, 7*(1), 115–120.

Varga, A., Kószó, M. F. Z., Mayer, M., & Sleurs, W. (2007). Developing teacher competences for education for sustainable development through reflection: The environment and school initiatives approach. *Journal of Education for Teaching, 33*(2), 241–256.

Vega-Marcote, P., & Varela-Losada, M. (2016). Basic teacher training oriented toward sustainability: Why and how to carry it out today? In W. Leal Filho & P. Pace (Eds.), *Teaching education for sustainable development at university level* (pp. 83–96). Dordrecht: Springer.

Vettori, O., & Rammel, C. (2014). Linking quality assurance and ESD: Towards a participative quality culture of sustainable development in higher education. In Z. Fadeeva, L. Galkute, C. Mader, & G. Scott (Eds.), *Sustainable development and quality assurance in higher education: Transformation of learning and society* (pp. 49–65). New York: Palgrave Macmillan. https://doi.org/10.1057/9781137459145.

Wi, A. (2019). Citizen participation as a key enabler for successful public education policies in climate change mitigation in Singapore. *International Research in Geographical and Environmental Education., 28*(1), 53–69. https://doi.org/10.1080/10382046.2018.143078.

Wibeck, V. (2014). Enhancing learning, communication and public engagement about climate change: Some lessons from recent literature. *Environmental Education Research, 20*(3), 387–411.

Wiek, A., Withycombe, L., & Redman, C. L. (2011). Key competencies in sustainability: A reference framework for academic program development. *Sustainability, 6*, 203–218.

Annie Hyokyong Nam is a graduate of the Technology, Innovation, Education Masters program at the Harvard Graduate School of Education (HGSE). She co-founded and led the first student environmental group at HGSE in addition to facilitating the first inter-school climate conference at Harvard University celebrating the 50th anniversary of Earth Day. She is currently working for the MIT Sloan School of Management and the Berkeley Haas School of Business, helping to digitize the program at both schools. Prior to Harvard, she worked for a private education firm, managing teachers and scaling core services by digitizing the curriculum. She has written a best-selling book on early English education for South Korean mothers.

Sueyoon Lee is a graduate of the Human Development and Psychology Masters program (Child Advocacy Strand) at the Harvard Graduate School of Education. She was a leader of the "Education and Communication" panel in the programming committee for the first inter-school climate conference at Harvard University. Her background centers around child rights advocacy. She volunteered for UNICEF Korea, planning and carrying out various campaign activities regarding child rights. She is certified as a Professional Instructor of Child Rights by the International Child Rights Center and has a certificate in Child Protection offered by the Harvard FXB Center for Health and Human Rights.

Chapter 7
Learning from Teaching Graduate Students How to Design Climate Change Education Programs

Fernando M. Reimers

7.1 Introduction

In this book I have presented the approach I developed to teaching climate change education in a professional school of education which consisted of engaging students in projects with educational institutions outside the academy, schools and non-formal education institutions, to assist those institutions in advancing efforts in climate change education. In a nutshell, this approach provides students an immersive experience of learning to do climate change education by doing it in a real world context infusing a project based experience into an existing course on policy analysis.

This pedagogical approach to teach climate change education produces three results: (a) an expansion in the capacity of schools and non-formal education institutions to develop climate change education programs appropriate to their context, as a result of the assistance provided by graduate students and reflected in the programs presented in this book, (b) university students who gain the skills to advance climate change education, and (c) an increase in the body of academic knowledge about climate change education as university students and faculty conceptualize and theorize this practice based work, as this book does. I think of these three products of this interlocking approach to climate change education as akin to the mathematically and nature inspired interlocking paintings of Dutch graphic artist Maurits Cornelis Escher (see https://mcescher.com/).

Notice that the approach differs from a direct instruction, didactic, approach of instructing students on the topic of climate change education, either into an existing course or in creating a course uniquely dedicated to the topic. Chapter 6 in this book, for instance, is an example of a course uniquely dedicated to the topic of climate change education. The latter has the obvious benefit of allowing greater

F. M. Reimers (✉)
Harvard Graduate School of Education, Harvard University, Cambridge, MA, USA
e-mail: Fernando_Reimers@harvard.edu

F. M. Reimers (ed.), *Education and Climate Change*, International Explorations in Outdoor and Environmental Education,
https://doi.org/10.1007/978-3-030-57927-2_7

opportunities to cover more relevant content but the inherent limitations that it requires dedicated faculty and students to the study of the topic, which may limit the number of students who enroll and the number of institutions who adopt the approach. The approach I have developed, which I see as complementary to courses focused exclusively on the topic of climate change which universities may also offer, is meant to make lower resource demands as a way to make it potentially more scalable across schools of education in universities around the world. To use an analogy to the sustainability curricula I developed for primary and secondary schools, which I discuss in the first chapter of this book, a course exclusively dedicated to climate change education would be the equivalent of the 'world course', whereas the approach I have proposed here, to infuse into an existing course on policy analysis an option that allows students to learn about climate change education as they partner with educational institutions is analogous to the second curriculum resource I developed in 'Empowering students to improve the world in sixty lessons' in which I created an approach to infuse sustainability curriculum widely throughout the entire school, including existing subjects, and offered five lessons per grade to be infused in those subjects. Given the much wider adoption of the 'sixty lessons' curriculum, I thought it reasonable to expect the same results in designing a similar approach for the tertiary education level.

The reason I see an approach that makes lower resource demands on institutions of higher education as necessary is because one of the challenges of advancing climate change education, as discussed in Chap. 1, is the challenge of doing this at scale. Climate change is progressing too rapidly to afford us the luxury of addressing it with small scale approaches to educating people in effective ways to engage them in adaptation, mitigation and reversal. I don't doubt the merits to have a small group of truly expert individuals, but experts alone are not enough to help either polar bears or humans survive the ravaging effects of climate change. We need to democratize access to effective climate change education and boutique solutions are inadequate to provide the scale such democratization demands. Frankly, this is the reason I underscore the urgency of infusing climate change education in primary and secondary schools, rather than stick only with universities as universities educate directly only a very small share of the world population, whereas most people now have access to compulsory education. This need to find solutions at scale is also the reason I have developed an approach which relies on university-school partnerships as they could become the largest and most widely distributed network of institutions globally to deliver climate change education. Collectively, universities have greater institutional capacity than any other institution I can envision could be deployed to support climate change education, and networked with primary and secondary schools, they can form an ecosystem of incomparable reach and effectiveness. The approach I offer is a path to the development of such an ecosystem.

There is an additional benefit of university school partnerships to advance climate change education, because universities are inherently more cosmopolitan than other education institutions, they can permit the collaboration among people in different geographies, from different walks of life, in this way enabling those participating to recognize the ethical considerations involved in determining burden sharing across generations, across countries and across groups with different levels of privilege.

In this chapter, I draw out seven themes from the analysis of the five curricula created by my students on education for climate change, and from reflection on the approach I created to introduce climate change education in the graduate course on education policy analysis in which these curricula were developed.

7.2 Educating to Address Climate Change Is About Active Problem Solving, Not Contemplation

In engaging with institutional partners to identify the way in which climate change is impacting specific communities and to develop ways to adapt to and mitigate climate change, my then students, and now former students and contributors to this book, learned not just to understand how climate change was impacting these communities. They also underwent a process of looking for various points of entry to make a difference, to help educate others on these issues. This way to learn about climate change education by doing, engaging in actual efforts to address climate change through education, provides richer opportunities to gain understanding of how climate change impacts communities that more detached and uninvolved approaches to study. I contrast this approach to learning about climate change education from engaging in the work from an approach I could have taken, I could have had the students read some of the reports of the International Panel on Climate Change and then, based on that reading, asked them to design a generic curriculum from the comfort of the Gutman Library, the library at the Harvard Graduate School of Education. This approach might have produced a valuable curriculum, and spared my students the additional time and complexities of having to negotiate a curriculum with existing institutions, but it would have been a generic product, specific to no particular place or population. Perhaps some institutions would have eventually found such curriculum useful, or they might have found it, as they did with my 'World Course' too detached from their particular goals, resources and constraints. I believe engaging my students first in understanding how climate change was affecting a particular place, and then in discerning how best to educate for climate change in dialogue with local actors, provided them a deeper learning opportunity, making the knowledge gained from the experience more personal than reading a report would have been and teaching them about the particulars of how climate change education can actually be advanced within the constrains and resources available to institutions, than designing a curriculum in a university library without the benefit of such interactions with the messiness of the real world.

The most obvious benefit of this pedagogy of learning by doing is that it can teach the students that the large, and seemingly intractable challenge of climate change writ large, can be tackled by translating it into smaller challenges, located in place and time, identifying specific populations, institutional partners, to address the challenge. In a nutshell, they learned to tackle a large and complex challenge by breaking it into smaller chunks. Because these students did this work not in isolation, but as part of a community of fellow students, which extended into ongoing collaborative work for an additional entire semester after the completion of the semester course in which they began this work, this helped them see their work as part of larger efforts, and made visible how the sum of small actions begins to aggregate into a larger, more impactful effort. Perhaps in the small collegiate community that collaborated in this book they will see a microcosm of the larger global movement necessary to address climate change. Most importantly, I hope this work that they produced, first in designing curricula and then in theorizing their work and communicating it to the scholarly and practice communities in climate change education, will help them see themselves as actors in that movement.

I realize that the curricula my students designed represent relatively small steps, in terms of the many actions necessary to produce systemic changes in the norms and institutions that undergird climate change or that produce sustainable development. Small steps, however, are necessary to help people find pathways that bring about more sustainable futures which will give them hope that change is possible. As Ojala has shown, too many people with knowledge about climate change are emotionally disengaged from the issue, overwhelmed by it, so it is essential to educate in a way that cultivates the right balance between worry and hope (Ojala 2012, 2016). What is learned from these small steps can provide the foundation for more complex understanding, and the efficacy gained from successfully completing these small steps can provide the confidence and the hope, to undertake more ambitious goals in the future. It is too easy for anyone contemplating the sheer complexity of the climate change challenge to drown in the complexity, to become paralyzed and hopeless. The pedagogical challenge, therefore, is to teach students that they can make a difference, and providing an experience at the right level of complexity, is a way to build their efficacy as well as their hope.

Mirroring what they had experienced, the same principle of learning not from contemplation but from action is expressed in the curricula designed by my students, most of which are project based efforts that propose to engage students in active efforts to address climate change in their communities.

One of the benefits of an action-oriented and action-based education approach to climate change is that it enhances the relevancy of what is learned. Learning to design climate change curriculum with strong connections to institutions in which such curriculum will be implemented has valuable self-correcting qualities that tame the risks of running too far afield in the imaginary worlds that are possible in ivory towers. In fact, reflecting on the deficiencies of many of the approaches to climate change education advocated by international development organizations I wonder whether they would have been more successful if those approaches had been developed with the benefit of strong connections to schools that had actually attempted to teach those curricula, and with subsequent opportunities for continuous improvement. Too many of the pronouncements of international organizations on what schools and educational institutions could do to advance climate change education reflect a very limited understanding of how schools and universities actually work and change.

The sheer choice of point of entry in the five projects that my students created illustrates the value of grounding learning about climate change education in engagement with practice: two of them focused on formal education institutions, two on non-formal environments and one on the potential role of schools of education supporting 'whole of system' change. This distribution of the foci of these projects resulted from the initial step all students had to follow of considering several alternative ways to address climate change through education in the particular location which was the focus of their work. In that process they considered formal and non-formal education, serving different student populations, establishing different partnerships. Their final choice was deliberate, and the product of analytic reasoning they had to articulate in the final paper they wrote for the course. It was part of the process of policy analysis they were learning in the course as they had to identify how specific impacts of climate change in that locality affected various human populations and to identify which of these populations had to be educated on climate change. As they focused on a particular population, students then had to identify the best pathway to reach that particular population in that jurisdiction. As a result, the choice of whether to focus on schools or non-formal institutions was not an arbitrary choice, made from the comfort of a library, but a choice resulting from engaging in an analysis of the context of practice in which they were grounding their evolving understanding of how education can support people in adapting to or mitigating the impact of climate change.

The resulting three approaches resulting from grounding the choice about these projects in practice differ from the distribution of approaches to climate change education reflected in two recent reviews of research in the field, which show a predominance of studies of school-based environmental education programs, with significantly more limited focus on non-formal environments (Jorgenson et al. 2019; Rousell and Cutter-Mackenzie-Knowles 2020), and an almost nil emphasis on teacher education institutions. A recent review of 221 studies of climate change education concludes, "Given the predominance of research in science education and environmental education, we were surprised to find the literature on teacher education to be relatively limited. Most teacher education studies we did locate focused on the climate change knowledge of pre-service teachers." (Rousell and Cutter-Mackenzie-Knowles 2020, p. 200). This lack of attention to non-formal environments and teacher education institutions are blind spots of the field, as effective education to address climate change requires more than work in schools and more than curriculum. This demonstrates that anchoring the design of climate change education in a process of analyzing various possible approaches to address the particular needs and opportunities in a specific context, in dialogue with local actors, produces a richer, more diverse, and perhaps more realistic, set of approaches than a top down approach that begins in the university library or even in the offices of development professionals of international organizations. This discipline of learning to collaborate with practitioners is valuable to help develop a perspective on the relevance of academic knowledge. A critique of research on environmental education argues that an important criterion to assess such scholarship is to ask whose ignorance does research reduce, proposing that too often the ignorance of researchers is not well aligned with the ignorance of potential users of the knowledge generated by research (Gough 2002, p. 19).

7.3 While Learning from Doing Is Valuable, to Advance the Field of Climate Change Education, it Is Necessary to Also Conceptualize and Theorize Practice

Two purposes are advanced as students conceptualize what they learned by doing as they engaged in a practice of designing a climate change education curriculum. The first is that theorizing practice advances the field of climate change education. The second is that conceptualizing practice contributes to professional preparation.

Given the contested nature of the field of climate change education, particularly whether it is a distinct field from environmental education and education for sustainable development, more research and theorizing are necessary to develop the knowledge base that would help the field mature. I argued in the introductory chapter of this book that there are two reasons to engage university faculty and students in the enterprise of supporting climate change education at precollegiate levels and in non-formal environments. The first is so that they can support the necessary development of capacity that the enterprise demands. The second is that universities can experiment, evaluate, conceptualize and theorize a field where, at present, practice leads theory, in order to transform it into a distinct field of practice guided by expert knowledge and supported by the powerful tools of logic and science.

For this reason, the engagement of students with a context of practice, and their work developing a climate change education curriculum, was only half of the work they had to complete in my course on education policy analysis. This engagement in practice had to be complemented by an analysis of their work, translating the knowledge they had gained into public knowledge, receiving feedback to their ideas in a conference and from academic peer reviewers, and finally publishing their work so it would become open to academic scrutiny and debate. In effect, the students engaged in a two-step of 'translation', first translating research in the field using it to inform the curricula of climate change education they created, and then translating back what they had learned from creating these curricula into academic scholarship which entered the academic conversation. Their chapters in this book reflect that second aspect of the work. If universities embraced the challenge of engaging more students, as part of their education, in similar activities developing approaches that can guide climate change education, we could accelerate the development of this field considerably.

Learning to conceptualize professional practice is also good professional preparation in general. In his book, *The Reflective Practitioner,* MIT Professor Donald Schon (1983), explained how one of the essential elements of professional preparation is to equip students with the skills and tools to reflect on their practice, and to draw from their practice knowledge which they can then make publicly available, falsifiable, supporting further professional practice as well as further research and scholarship. This ability to reflect on the knowledge which guides practice is essential to the improvement of professional practice. Practitioners often guide their practice with knowledge to solve problems that goes beyond the mechanic application of principles or conclusions drawn from academic knowledge. Schon also argued that the failure to comprehend this all too often leads institutions involved in professional education to base the curriculum on a paradigm which assumes that professional practice is simply the application to problems of practice of the general principles drawn from academic research in the fields associated with that practice.

Schon's epistemological stance recognizes that when practitioners solve problems they learn from the consequences of their actions, and the knowledge they gain makes them better at solving problems in the future, hence better professionals. Solving problems, especially complex, messy, adaptive or divergent problems such as climate change, is thus much more than mechanically applying lessons drawn from research to new situations, and involves forms of creation, design of solutions and experimentation. While good professionals learn from these 'private' experiments that constitute their practice, this knowledge is often accessible only to the practitioner, because it is not processed in a way that allows others to learn from it or to falsify it. This is called 'tacit' knowledge. Constructing opportunities to learn from such knowledge, making 'tacit' knowledge 'public' is thus critical to professional education, and to the advancement of the profession.

Some of the critiques to professional education based in universities, concern whether the curriculum provides sufficient access to knowledge essential for effective practice, or whether it is too 'theoretical' too 'disconnected' from the fields of practice for which they are preparing people for professional practice. In *The Reflective Practitioner* Schon argues that the classical worldview that sees practice as a mere application of foundational principles is responsible for this disconnect. The approach I developed, which engages students first in designing an educational program in dialogue with partners in institutions of practice, and then in examining the results of their work and discussing it in the context of the larger academic scholarship on climate change education attempts to close the disconnect described by Schon.

7.4 What Outcomes Matter in Climate Change Education

The key learning outcomes that I sought to help my students gain were the ability to understand the interrelations between climate change and specific populations, to identify knowledge, attitudes and skills that would help those populations adapt to and mitigate climate change, to develop an actionable approach to climate change education in those communities that would help those population gain such competencies, anchored in a robust theory of action and in a feasible strategy of implementation, with adequate local support to ensure sustainability. I further sought to help them reflect on what they had learned, and conceptualize and communicate their efforts in ways which made the knowledge they had gained visible and public, so that their experiments represented a contribution to advancing the field of climate change education. Knowing a topic, no matter at what level of depth, is only one component of the competency to deploy this knowledge in service of transformative goals, such as supporting school change.

In a nutshell, I sought to help my students break down a complex problem such as how to educate about climate change, into a tractable, smaller, problem, and to design a solution that could be tested in a particular social context. In this way, I hoped to help them become aware of their own power to make a difference, individually and collectively, in changing systems that undergird climate change education. The power and the desire to personally engage, the competency to address climate change education, requires more than knowledge of the research on climate change education or the research on climate change. It requires the cultivation of self-knowledge, of ethics, of empathy, or hope, or skills to collaborate with others. That breadth of skills is what a project based immersive experience in having to solve a real problem of practice helps develop in ways which simply reading reports in the library about climate change or research on climate change education do not.

So far, the designs my students have produced, and these chapters, are testimony of what they learned from this experiment. The impact of this book in the field, and my former students eventual continued engagement with the efforts they began in this course and with climate change education more generally, will provide evidence in years to come to evaluate some of my hypotheses on the long term outcomes of this form of professional preparation.

Specifically, I sought to help my students gain the skills to develop a context specific strategy for climate change which addressed these questions:

1. What are the specific impacts of climate change in this jurisdiction? How do they impact various human populations? Which of these populations needs to be educated on climate change?
2. What knowledge, dispositions and behaviors could mitigate the impact of climate change and are there ways in which changes in the behaviors of populations in this jurisdiction could slow down climate change?
3. What are the means of delivery to reach each of the specific populations in this jurisdiction?
4. What curriculum can best educate each population?
5. What institutions can support the development of the institutional capacity necessary to deliver such curriculum effectively?
6. What institutional collaborations can support the implementation of this strategy?

From this analysis, I sought to help my students develop and assess climate change education programs grounded in sound logic theories, where they could make explicit the hypotheses which undergird any climate change curriculum or instructional program. I also sought to help them identify and assess, weighing various criteria, alternative paths to address the root causes that undergird the lack of knowledge, skills and attitudes that prevented the populations they were serving from adapting or mitigating climate change.

The chapters in this book demonstrate that students now have those skills. In these chapters, my former students demonstrate that they competently used policy analysis approaches to assess alternatives to climate change education, and to develop a sound logic theory to sustain their program, addressing the six core questions.

I see less evidence, however, that I succeeded in helping my students construe the implementation of such a curriculum as an adaptive challenge that would require attending to the cultural, psychological, professional, institutional and political dimensions of the enterprise (Reimers 2020).

Relatedly, I see only incipient evidence that the curricula designed by my students aim to help those who engage with that curricula develop an understanding of the systems that undergird climate change. This emphasis in understanding systems is most evident in Chaps. 2 and 3, the school-based curricula in the Middle East and Guatemala. I was aware that the opportunity to gain the skills to understand and transform systems eludes most of the efforts of climate change education examined in a recent review of research (Jorgenson et al. 2019) and had hoped that grounding an approach to climate change education in engaging students with local institutions and local actors, would seamlessly translate into understanding the systems which undergird current climate education challenges, and to discern what kind of collective action would be involved in changing them. I assumed that such engagement would translate in understanding changing education institutions as an adaptive challenge requiring a multi-stakeholder coalition that could produce collective leadership to sustain the change effort.

Perhaps the reason I did not achieve this goal stems from the inherent limitations of conducting this work within the timeframe of an academic semester in the context of one of the multiple courses students are taking. It is possible that a more intense immersive experience, for instance working solely on this project for an entire semester, and using that work as the anchor for inter-disciplinary integration of all other academic study in their graduate program, might have produce deeper learning. A different engagement, in which students could sustain partnerships with schools through various cycles of implementation, revision and adaptation of the curriculum, might also have made more visible to them the 'systems' which have to be changed in order to institutionalize their climate change curriculum. While I hope that they will continue such engagement with the institutions that they partnered with, that work will now take place outside the context of the course. It is also possible that I should have relied on more didactic approaches to teach some of this knowledges, rather than expect that it would be an inevitable by-product of the experiential learning I was engaging my students in. There are other limitations resulting from carrying out this kind of action based climate change education during a semester long course, we overcame them in part because all students continued to work on these projects after the semester had ended, particularly making revision to their chapters and responding to the various rounds of feedback they received. While it is satisfying to know that there is enough interest in learning for its own sake so that a group of dedicated students can commit to work an entire additional semester, after they had completed the course on which this work began, on a project of this sort when this does not fulfill any requirement or has any grade attached to it, I wonder if this approach is scalable. An optimistic interpretation is that if a learning experience truly engages students, that engagement can sustain continued learning even when the original structures that supported it, the course requirements and the academic credit, are no longer there.

It is also possible that there are limitations to how much can be expected from an infusion approach to helping students learn about climate change education in the context of existing courses. As I explained earlier in this chapter, these limitations of the approach I followed are also its virtues. While the engagement in practice this semester-long, turned into year-long, assignment provided students was somewhat limited, perhaps not sufficiently intense, or long or immersive, that is also why I was able to fit it into my course, and students were able to fit it into their schedules which included other courses in addition to the one they were taking with me. If I compare the approach I developed and implemented in my course with the K-12 sustainability curricula I had developed, what I did was more similar to the 60 lessons curriculum than to the original ‘world course’. Recall that the ‘world course’ was an intense and immersive experience, engaging students in a deep, coherent and rigorous sequence of 350 units, requiring 10 h a week from kindergarten to high school. This was, perhaps, too much of a good thing, too demanding for most teachers to be able to implement in their schools. Translating this ‘heavy’ approach to sustainability education into a ‘lighter’ 60 lessons curriculum, five lessons per grade, made the approach a lot more usable and scalable. My hope is that the approach I have developed and describe in this book, to create opportunities to develop climate change education curricula infusing them into existing courses, is more scalable, than proposing an immersive climate change education course, the parallel to the ‘world course’, that students should take. Certainly, students may learn more about climate change in that dedicated course, but it is likely that such an approach would reach fewer students than infusing climate change related assignments in a wider array of existing courses.

Similar to my learning goals for my students, the learning goals of the curricula which they designed espoused similarly ambitious goals for the students they sought to serve. Chapter 2, the curriculum to cultivate climate change leaders in the Middle East, seeks to foster leaders who can engage in civic action around climate change, and who are capable of practicing empathy, systems-thinking, media literacy and collaboration. The curriculum that Margaret Wang and David Rhodes designed included not only an essential foundation in the sciences, but critical skills to interpret scientific information, along with agency gained in a hands-on project that teach students the role of complex power structures, systems and the needs and interests of various stakeholders. The curriculum aimed at teaching students how to engage with climate change as individuals but also collectively to affect climate change.

Chapter 3, the curriculum of whole school climate change in Guatemala focuses on helping students learn to think about systems, understand climate change as a shared responsibility in order to change systems.

Focusing on marginalized populations, Chaps. 4 and 5 focus on non-formal education approaches. Chapter 4 focuses specifically on how to mitigate the impact of climate change among vulnerable populations in Haiti using radio education to develop knowledge, behavior and skills that minimize the harm caused by climate-related disasters and to help them adopt sustainable agricultural methods. Chapter 5 focuses on out of school youth in Pakistan to develop social and collaborative skills that help them work in teams and develop their own solutions to mitigate the impact of climate change in their lives and to discover their voice in advocating for their rights. In Natasha Japanwala's words: "it is important to teach students how to advocate for themselves — not just so that they can demand better resources and negotiate if they have to migrate, but because they should be able to conduct public protests or speak to local authorities if need be."

Chapter 6, in proposing student led curriculum about climate change in schools of education, advocates tapping the power of students to transform the contexts in which they are learning, essentially cultivating their agency and efficacy in transforming systems as a way to contribute to a grassroots movement of climate change advocates.

7.5 The Power of Contextually Situated Learning

The pedagogy I followed in teaching my students about climate change invited them to find a specific context on which to ground the practice they would engage through the course and beyond. I actively discouraged students from writing 'generic' guides or curriculum of climate change education, and instead urged them to identify a specific site and institutional partners so their practice could be constructed in dialogue with those partners. My main goal was to ground this work on dialogue with local actors that would cultivate deep empathy with specific communities and institutions, and help my students learn how local populations articulated the impact of climate on their lives and the role of education in helping address this impact. This helped students investigate the specific ways in which climate change was impacting particular communities, and from there, discern what kind of learning outcomes would be of greatest value to those particular communities. This crucial step is, by necessity, missing from any generic guide on education about climate change designed to serve a multiplicity of contexts.

Predictably, once my students focused on the specific learning outcomes that would be of value to the students participating in the programs they were designing to cope with or mitigate climate change, the education programs they designed focused also on those specific manifestations of climate change. The leadership curriculum presented in Chap. 2, for instance, engages students in observing environmental degradation in their particular communities, and in developing causal models and networks of persons or institutions implicated in the process, using this to teach students systems thinking.

In designing a whole school approach to climate change for Guatemala, Chap. 3 analyzes both the limitations of UNESCO's generic guide of whole school change, as well as the limitations of Guatemala's curriculum on climate change, developing an approach grounded on the particular strengths and limitations of a particular school. Their analysis of the shortcomings of the national curriculum calls out "the overall lack of interdisciplinarity within the Base National Curriculum, and of teacher professional development opportunities geared towards building competencies on environmental and climate topics, indicate a lack of acknowledgement among the country's educational authorities about the complex and interdisciplinary nature of this issue."

Chapter 4, proposing a radio education program to educate Haitian adults to mitigate the impact of hurricanes in their lives, recognizes the shortcomings of the formal education system in excluding many students and in leaving many graduates very poorly educated. It is this analysis of context which leads the authors to focus on adult education as an essential avenue to address the impact of climate change on vulnerable communities.

Similarly, Chap. 5 recognizes that addressing the impact of climate change on the most vulnerable requires focusing on those who are poorly served or excluded from the formal education system. Focusing on a particular population of out of school vulnerable youth in southern Pakistan, the authors of the chapter integrate a climate change education curriculum within a life skills and literacy skills curriculum. This contextual grounding guides also the very content and structure of the curriculum:

> The first step of the curriculum is helping students understand what climate change actually is, rather than instructing them about climate change as it is occurring around the globe, the introductory unit of the curriculum focuses on three phenomena that impact Badin directly and that the students will have observed around them: overheating in the summer, floods because of sea levels rising, and the lack of productivity of the land because of salinization. This is largely in line with the overarching philosophy driving climate change curricula: for example, the US-based National Center for Science Education (2016) describes four best practices for climate change education that can be considered universal: "Make it local, make it human, make it pervasive, make it hopeful."

A distinctive feature resulting from grounding these projects in specific contexts visible in the two programs designed for formal schools is that they are integrated with existing curriculum in use in those institutions. The leadership curriculum in the Middle East builds on an analysis of ongoing curriculum efforts to teach about the environment and climate, and relates the proposed curriculum to them. The whole school approach to climate change proposed for Guatemala is also anchored in an analysis of the current curriculum in Guatemala.

Chapter 6 illustrates the power and the limitations of contextually situated learning. The goal of the chapter is to formulate an approach that can support education for climate change by educating broadly students that prepare for various professional roles in education, the 'proof of concept' is developed in the specific context of the Harvard Graduate School of Education. While there are similarities across schools of education, there are also particularities in how they are governed, for example in how open they would be to having students propose and lead a course on climate change. This inherent limitation of developing a case deeply rooted in a unique context is shared by the other projects presented in this book.

Another benefit of grounding this pedagogy of learning to address climate change in local contexts is that it helped students engage with local actors, and develop the skills for collaboration that will arguably prepare them to continue to advance the necessary systemic change to curb climate change.

7.6 A Pedagogy to Change Climate Through Education

In my course on education policy analysis I adopted a problem- and project-based pedagogy designed to cultivate the agency and problem-solving skills of my students engaging them in co-constructing the curriculum they were experiencing. Essentially, I offered four alternative paths to develop the skills in policy analysis which were the subject of the course, one of these paths was designing a climate change curriculum. I then offered guidance on how to find a context and which questions to address, meeting with those teams as necessary, but for the most part letting them do this work independently. The support the students received came in the form of readings such as the reports of the International Panel on Climate Change and some existing curricula on climate change and readings on climate change education, sharing my own experience designing sustainability curriculum and working with schools and school networks around the word. The students were supported more generally in the course with other readings on system level change, curriculum and teacher education, that were the basis of lectures, classes and discussions in this course, but those focused on comparative education, not on climate change education in particular. Students did receive feedback to various papers they wrote, all of them building up to the final paper in the course, and of course they had the benefit of a community of peers working on similar topics. But the intellectual autonomy and freedom that undergird my pedagogy was closer to that which is typical of writing a research paper, or a dissertation, than that which is typical of a scripted graduate course in which students have to complete pre-specified assignments or tests which are then graded on a rubric.

Unsurprisingly, but to my pleasure, my students followed similar pedagogical principles in the curricula which they designed. Whether this is because this assignment in my course attracted students with a penchant for intellectual autonomy, or because experiencing it caused them to seek similar freedom for the students they were serving is something I am not able to determine. Chapter 2 proposes a student centered, project-based curriculum in the Middle East, as does Chap. 5 in Pakistan. Chapter 6 takes this idea even further, by proposing not a student co-constructed curriculum, but a student led, and faculty supported curriculum in the very same institution in which I teach. Only the adult education curriculum via radio-education in Haiti adopts a didactic approach with a specific focus on helping adults develop knowledge, attitudes and skills that can help them minimize the impact of hurricanes on their lives. The emphasis adopted by most students on project-based pedagogies resonates with the conclusion of a recent review of climate change education programs, which identifies a gap in the existing literature in terms of approaches to climate change education which are participatory, interdisciplinary, creative, and affect-driven (Rousell and Cutter-Mackenzie-Knowles 2020, p. 191).

> We found that the four approaches which have dominated the literature on climate change education were generally top-down approaches, whether the focus was on scientific knowledge, formal curriculum, behaviour change, or mitigation/adaptation. Yet underneath this entrenched edifice of top-down education and disaster management, a series of innovative,

> bottom-up approaches have begun to emerge. These include participatory approaches which empower communities of learners to design their own climate change projects and modes of engagement with the issue. (Rousell and Cutter-Mackenzie-Knowles 2020, p. 202).

> More specifically, this review identifies a pressing need for research that gives young people both a hand and a voice in redressing the complex implications of climate change in their own communities and environments. Our analysis calls for new ways of making climate change meaningful for children and young people through participatory and arts-based modes of engagement. In effect this is extending climate change education and its research beyond the realms of understanding young people's climate change science knowledge (or lack thereof), which has no bearing on climate change itself, to far more important and pressing aims which actively empower children and young people to mitigate climate change. (Rousell and Cutter-Mackenzie-Knowles 2020, p. 203).

The non-formal education curriculum for out of school youth in Pakistan articulates the importance of this type of pedagogy to the critical outcomes it seeks to develop. As Natasha Japanwala described,

> Using design thinking and project-based learning to teach adaptation strategies has value beyond familiarizing the students with new skills and the ability to work collaboratively towards a common goal. It builds their leadership skills by encouraging them to think of themselves as climate change ambassadors in their community, especially since the exercises follow a unit where their conceptual knowledge was built. They have, through units 1 and 2, had both the experience of leadership and the vocabulary to articulate the cause they are leading.

The curriculum I developed was a climate change education strand embedded in a comparative education policy course in which students would work on projects to support schools or other education organizations, one of four options students could pursue to practice their policy analysis skills. As such, my course was not interdisciplinary but very much a course on education policy and comparative education – although comparative education is an interdisciplinary field. We drew on knowledge on teaching and learning, education system change, and policy analysis. Obviously additional knowledge informed the work of my students, but this was knowledge I assumed they would bring in, or gain independently as they worked on their projects. In other words, I thought of the action projects in which the course engaged them as the occasion that would help students generate questions which would lead them to access the bodies of knowledge necessary to solve the problem. This was especially the case in investigating the way in which climate would affect the particular populations and localities of their projects, an area on which I played no role in instructing them, other than suggesting resources they could consult. All learning in this area was self-initiated and self-directed.

This is one area in which the curricula my students designed does not mirror the curriculum they experienced in the course. Most of them did propose interdisciplinary approaches. Perhaps this is because there are certain affordances and assumptions one can make in teaching Harvard graduate students which are less feasible when teaching at lower levels of education. In particular, the two curricula designed for the formal education system, in the Middle East and in Guatemala, were interdisciplinary. With respect to the two curricula designed for vulnerable out of school population the curricula were interdisciplinary too, although the implementation vehicle for those curricula was not structured around disciplines and subjects in the same way schools are. The student led curriculum for the Harvard graduate school of education presented in Chap. 6 is also interdisciplinary, although it benefits from the same affordances my course did, on relying on background knowledge of students and on their ability to learn what they need to work on their personal projects.

7.7 Augmenting the Capacity for Climate Change Education Among Teachers and Schools

One of the explicit aims I sought in urging my students to think about climate change in specific contexts and with concrete institutions in mind was to help them consider institutional capacity in their efforts to address climate change education. As I explain in the introductory chapter of this book, it is the lack of attention to institutional capacity which accounts for the insufficient results of past efforts in climate change education. This theme is also a central tenet of the course on comparative education policy analysis where this work on climate change was carried out. The readings in the course support the proposition that policies are not self-executing and that good policy analysis considers implementation as part of the process of analysis itself, and not as an afterthought. My students' designs demonstrate that they understand the necessity to think about capacity to implement a climate education program, and that any climate change curriculum must achieve a balance between being within reach of the existing institutional capacity, while creating the opportunities to strengthen that capacity. All chapters in this book identify institutional partnerships as avenues to strengthen the institutional capacity of teachers and schools and propose integrating programs of professional support to help teachers develop their capacity.

Chapter 2, the leadership curriculum for the Middle East, argues the value of a partnership with a higher education institute, the Arava Institute, as a way to strengthen the capacity of teachers to teach about climate change. It also proposes an iterative process of improvement in which teachers collaborate with others in professional communities as a way to develop their capacity to teach about climate change. What is very clear in the programs presented in this book is that strengthening institutional capacity is not an afterthought to the design of a climate change education program, but an integral part to the design of the program.

Chapter 3, identifies teacher professional development as a crucial omission in UNESCO's guide on whole school climate change education (Gibb 2016), and proposes an approach to build that capacity in the school that is the focus of their project through a partnership with a local environmental education 'Amigos del Lago'.

The two projects focusing on nonformal education include an analysis of possible institutional partners with the necessary capacity to execute the programs.

Chapter 4, focusing on radio education, maps a network of institutional partners with the know how to design and deliver high quality radio programs.

Chapter 5, embeds the program of climate change education into the programming of an institution working to develop literacy and life skills among the target population for the program, the Citizen Foundation, and recognizing the financial limitations of the organization, is explicitly designed to be low cost and low resources.

Chapter 6, adopts an entrepreneurial approach to the question of capacity. Recognizing the limited existing capacity in schools of education to educate for climate change, it proposes to mobilize an underutilized resource, the students, and makes them the agent of curriculum renewal. This potentially transformative idea assumes, of course, the leadership capacity to embrace this role among students.

7.8 Blind Spots

A recent review of research on climate change education argues that addressing climate change education requires more than changes in individual consumption, but collective action that can change systems. That review identifies as a limitation of the field that much of the research, and of the practice, in climate change education is based on a mindset focusing on individual impact on climate change (Jorgenson et al. 2019, p. 160). The authors of the review argue that reducing the risks of climate change requires a sociotechnical transition towards more renewable-energy systems which would call for complex social, technological and ecological changes. In their view, a relevant climate change education needs to help students understand and bring about those changes. At present, the reviewers argue, most environmental educators and researchers are focused on influencing individual behavior, thus privatizing the concept of environmental action.

> By minimizing the role of collective action, environmental educators and researchers may be reinforcing a simplistic and narrow conception of the relationship between climate change, human action, and energy system change and distorting the fact that many of the most impactful climate actions are decisions about energy supply systems that are made by state and market sector actors under direct pressure from advocacy coalitions and other social collectives. (Jorgenson et al. 2019, p. 166).

The five projects created by the students in my course follow within the more conventional paradigm focusing on influencing individual behavior rather than on collective action across diverse stakeholders and institutions that can help bring about the sociotechnical transition towards renewable energy systems. While some of the projects, such as the leadership project in the Middle East or the non-formal project in Pakistan, emphasize the role of collaboration and collective action, the scope of that construct is limited to collective action among students or adult learners in the context of school projects or literacy and life-skills programs. It does not contemplate educating for collective action across institutions and diverse stakeholders and lacks explicit connections to macro-level transformations of energy systems. This could be the result of focusing on schools or non-formal education institutions as the point of delivery of the curriculum, as it is a conventional way to see schools as working in some isolation from other institutions. While several of these projects propose partnerships between schools and other institutions, for example the projects in the Middle East and Guatemala, these are generally for the purpose of enhancing the capacity of schools to teach about climate change. There are obvious limitations in the scope of a one semester course to engage students in partnerships with schools that do much more than work within the boundaries that schools accept as legitimate for their work. The notion that more effective environmental education would require creating opportunities for students to collaborate with multi-actor networks, such as climate scientists and activists, renewable energy firms and entrepreneurs, state agencies, NGOs and civic groups, is probably outside the scope of what a group of students can realistically help orchestrate in collaborating schools over the course of a semester.

A seeming logical byproduct of developing climate change education programs which are contextually situated does not seem to have materialized in these five projects. One could expect that any design for specific contexts would take note of existing knowledge and beliefs about climate change. The importance of designing climate change education taking existing beliefs, attitudes and contexts of learners into account, rather than assuming they are blank slates is underscored in a critique on how current approaches to climate change curriculum shape attitudes and behaviors towards climate change (Brownlee et al. 2013). In none of the curricula presented in this book is there a recognition of what those pre-existing beliefs and attitudes are in the specific contexts for which these curricula were developed.

The curricula developed by my students demonstrate only an emerging understanding of the interdependence of climate change with wider processes such as poverty alleviation and sustainable development. While most of these curricula address the impact of climate change specifically on people in poverty, and four of the five are focused on the developing world, there is no specific recognition of the gendered experience of climate change.

Finally, a word about a tension in a project which began with the premise that in order to be more effective, climate change education efforts would need to be well grounded in specific local contexts. As I argue in the introduction of this book, there are limitations to curricula or instructional materials that are written at such level of generality that the ignore everything that is defining of any particular context. This is not to say that every program of climate change education needs to be developed from scratch for every particular context. To make such radical case for the need of idiosyncratic climate change education would be to negate the possibility of building a generalizable corpus of knowledge that can inform the enterprise. I am not making such case. But I am suggesting that it is important to add to the corpus of knowledge that can help make climate change education a distinct field of study and practice, knowledge that is the result of particular efforts to transform practice. Each of those efforts can be considered an experiment of sorts. The reports of such experiments need to be conceptualized so that subsequent reviews can identify generalizable principles from them. We obviously hope, in writing this book, that what we learned in these five settings would be of value to other climate change educators in different settings, and of value to those interested in theorizing the field. I am also hopeful that while this book reports on a pedagogical experiment on climate change education in a single course at the Harvard Graduate School of Education, the analysis offered here would be of value to colleagues in other institutions working in schools of education and perhaps even in other fields such as public health, public policy, government or business. I have offered the premise that if more universities engaged with this work, this would help fill an important gap in institutional capacity that is at the core of the limited effectiveness of present climate change education efforts. This premise can only be tested if a sufficient number of colleagues in other institutions, and even in my own, accepts the invitation to give this a try, and this book is meant to suggest that it is at least worth trying.

7.9 Coda: Writing About the Role of Universities in Climate Change in Education During a Pandemic

I review these concluding lines in mid september of 2020, six months since Massachusetts public health authorities instituted various physical distancing measures to contain the spread of the COVID-19 Pandemic. My course last semester shifted to online instruction on short notice, as did the instruction of most educators around the world. We have recently started the academic semester online, in what will be the first ever entire academic year in which all curriculum will be delivered online at the Harvard Graduate School of Education. Globally, at the time of writing, more than twenty nine million people have been diagnosed as infected and more than 929,000 people have died from COVID-19. I suspect many more will still be infected and die until an effective treatment is available. The prospects of developing and distributing a vaccine to prevent infection or of discovering and effective treatment to mitigate the lethality of infection are at best six to twelve months away. In addition to the direct impacts of the pandemic on health and life, other indirect effects ravage the planet: unemployment, hunger, poverty, rising authoritarianism. I certainly hope this pandemic does not lead also, as did the Flu Pandemic of 1918 in Germany, to the breakdown of democracy and the rise of fascism (Blickle 2020). Underscoring the severity of the crisis created by the Pandemic, Antonio Guterres, Secretary General of the United Nations, has characterized it in this way "The Pandemic is more than a health crisis, a security crisis, and a human rights crisis. It has affected us as individuals, as families and as societies. The crisis has highlighted fragilities within and among nations. It is no exaggeration to suggest that our response will involve remaking the very structures of societies and the ways in which countries cooperate for the common good. Coming out of this crisis will require a whole-of-society, whole-of-government and whole-of-the-world approach driven by compassion and solidarity." (un.org/en/coronavirus/UN-response)

I have focused the last six months in assessing the educational needs and the education responses to the pandemic around the world, and developing tools which can help education leaders develop strategies for education continuity. As I was concluding this book during this period of physical distancing and living and working online, I see in this pandemic and in our inadequate education responses to it some parallels to the topic of climate change and our lack of adequate education responses.

In most countries, the education response to COVID-19 was slow, reactive and short term, rather than proactive and long term (Reimers and Schleicher 2020a, b). As a result, there are vast differences across schools and jurisdictions in how children from different circumstances have been impacted by this pandemic. Some have benefited from sustained opportunities to learn, through various means. Others, not so much. As a result, learning gaps are increasing and it is likely, because further periods of physical distancing are to be expected until there is a vaccine to prevent infection or an effective treatment to reduce the health impact of infection, that learning loss resulting from these gaps in learning opportunities will continue. If they do, this pandemic could cause the greatest global loss in opportunity to learn in decades, if not centuries.

Such massive education losses were not the inevitable result of a pandemic, and neither were the number of people who have been infected or who have died. The health impact of the pandemic has been mediated by the institutional capacity and leadership effectiveness in various jurisdictions. At present, the reported number of COVID-19 related deaths per capita varies widely across the world, alongside the preparedness and efficacy of the response of public health and political authorities, and the ways in which the data is recorded and reported. In some jurisdictions, leaders paid attention to the evidence, took the advice of public health authorities, and acted decisively and effectively. In other jurisdictions, not so much. As a result, the number of deaths per 100,000 people ranges from 96 in Peru, 65 in Bolivia, 64 in Chile, 59 in the United States, to 6 in Finland, and even less in India, Poland, Egypt and a number of other nations (Johns Hopkins University 2020). As time goes by, and we obtain more accurate estimates of number of actual deaths related to COVID-19, which are probably underestimated at the moment given deficient diagnostic measures, and as the pandemic runs its course throughout the world, perhaps in several waves of infection, we will have a better sense of the extent of these differences.

Similarly, the education effects of the pandemic are also mediated by the institutional capacity and leadership effectiveness of various jurisdictions, even of different schools. Where there was readiness to sustain education remotely, where there were established alternative delivery mechanisms, such as connectivity and devices to reach students, and where students had skills for independent learning, there has been more educational continuity than where such alternative delivery channels or skills for independent learning were missing.

As a result of the deficient education preparedness and response, as countries have enacted regimes of physical distancing, millions of students around the world are stranded like polar bears as their learning platforms have melted, watching their learning opportunities vanish.

COVID-19 is a high impact event just like climate change, only it is happening much faster. Perhaps it is the sudden and rapid nature of the spread of this highly contagious virus that impeded our capacity to adapt our educational institutions so they could continue sustaining learning under different conditions to help us cope with this massive change to the way we live. However, since the early days when the World Health Organization declared the pandemic, various models projecting the rate of infection of COVID-19 could have helped inform what actions could be put in place to slow down the spread of the virus. Some jurisdictions made better use of that information than others, with consequential results for the lethality of the pandemic across jurisdictions.

Climate change is a high impact event as predictable as a pandemic. The various effects of Climate change can be predicted with a particular likelihood of certainty, and the reports of the International Panel on Climate Change provide those estimates (IPCC 2018). From those estimates we know that many of these effects of Climate Change are much more likely than a Pandemic. Perhaps happening more slowly, giving us more time to respond. As with the models predicting the likely course of the pandemic, we have sufficient information now to model and anticipate how climate change will continue to unfold. We could make use of this information to inform individual and collective actions that will have consequential results for the lethality of climate change.

Those actions require that we build human capacities and the motivation and the skill to affect those individual and collective changes. Climate change education is a field of scholarship and practice that can guide the efforts to build those capacities. Will we be able to act more swiftly and more effectively than we have been to in the case of the pandemic? Will the pandemic have made sufficiently visible for us our footprint on the environment, to cause us to want to slow down climate change? Will we be able to generalize what we have learned about the devastating consequences of a high impact global calamity like the Pandemic and about the consequences of acting effectively or of failing to act to the predicted high impact of Climate Change?

My hope is that university students, and faculty, will see that addressing these questions is core to their mission even, and perhaps especially, as we ponder how this pandemic will change us all and our way of life and how we gear up to build back better in the Pandemic's aftermath.

References

Blickle, C. (2020, May). *Pandemics change cities: Municipal spending and voter extremism in Germany, 1918–1933*. Report No. 921. New York: Federal Reserve Bank of New York. https://www.newyorkfed.org/research/staff_reports/sr921

Brownlee, M. T. J., Powell, R. B., & Hallo, J. C. (2013). A review of the foundational processes that influence beliefs in climate change: Opportunities for environmental education research. *Environmental Education Research, 19*(1), 1–20.

Gibb, N. (2016). *Getting climate-ready: A guide for schools on climate action*. Paris: UNESCO. https://unesdoc.unesco.org/ark:/48223/pf0000246740.

Gough, N. (2002). Ignorance in environmental education research. *Australian Journal of Environmental Education, 18*(2002), 19–26.

IPCC. (2018). Summary for policymakers. In V. Masson-Delmotte, P. Zhai, H.-O. Pörtner, D. Roberts, J. Skea, P. R. Shukla, A. Pirani, W. Moufouma-Okia, C. Péan, R. Pidcock, S. Connors, J. B. R. Matthews, Y. Chen, X. Zhou, M. I. Gomis, E. Lonnoy, T. Maycock,

M. Tignor, & T. Waterfield (Eds.), *Global warming of 1.5°C*. An IPCC special report on the impacts of global warming of 1.5°C above pre-industrial levels and related global greenhouse gas emission pathways, in the context of strengthening the global response to the threat of climate change, sustainable development, and efforts to eradicate poverty. In Press.

Johns Hopkins University. (2020). Coronavirus Resource Center. Mortality Analyses. https://coronavirus.jhu.edu/data/mortality. Accessed 11 May 2020.

Jorgenson, S. N., Stephens, J. C., & White, B. (2019). Environmental education in transition: A critical review of recent research on climate change and energy education. *The Journal of Environmental Education, 50*(3), 160–171.

Ojala, M. (2012, October). Hope and climate change: The importance of hope for environmental engagement among young people. *Environmental Education Research, 18*(5), 625–642.

Ojala, M. (2016). Facing anxiety in climate change education: From therapeutic practice to hopeful transgressive learning. *Canadian Journal of Environmental Education, 21*, 41–56.

Reimers, F. (2020). *Educating Students to Improve the World.* Singapore: Springer.

Reimers, F., & Schleicher, A. (2020a). *A framework to guide an education response to the COVID-19 Pandemic of 2020*. Paris: OECD.

Reimers, F., & Schleicher, A. (2020b). *Schooling disrupted, schooling rethought. How the COVID-19 Pandemic is changing education*. Paris: OECD.

Rousell, D., & Cutter-Mackenzie-Knowles, A. (2020). A systematic review of climate change education: Giving children and young people a 'voice' and a 'hand' in redressing climate change. *Children's Geographies, 18*(2), 191–208.

Schon, D. (1983). *The reflective practitioner. How professionals think in action*. New York: Basic Books.

Fernando M. Reimers is the Ford Foundation Professor of the Practice of International Education and Director of the Global Education Innovation Initiative and of the International Education Policy Masters Program at Harvard University. An expert in the field of Global Education, his research and teaching focus on understanding how to educate children and youth so they can thrive in the twenty-first century. He is a member of UNESCO's Commission on the Futures of Education.

He has written or edited over a hundred academic articles and chapters and thirty-four books, of which the most recent, focusing on system level change to make education more relevant, include: Implementing Deeper Learning and 21st Education Reforms. Building an Education Renaissance after the COVID-19 Pandemic; Educating Students to Improve the World; Audacious Education Purposes. How governments transform the goals of education systems; Empowering teachers to build a better world. How six nations support teachers for twenty-first century education; Letters to a New Minister of Education; Teaching and Learning for the 21st Century; Preparing Teachers to Educate Whole Students: An International Comparative Study.

CPSIA information can be obtained
at www.ICGtesting.com
Printed in the USA
BVHW040156301220
596723BV00005B/27

9 783030 579265